T0269521

How to Scale-Up a Wet Granulation End Point Scientifically

How to Scale-Up a Wet Granulation End Point Scientifically

Michael Levin, PhD
Milev, LLC
Pharmaceutical Technology Consulting
Measurement Control Corporation (MCC)

AMSTERDAM • BOSTON • HEIDELBERG • LONDON
NEW YORK • OXFORD • PARIS • SAN DIEGO
SAN FRANCISCO • SINGAPORE • SYDNEY • TOKYO

Academic Press is an imprint of Elsevier

Academic Press is an imprint of Elsevier
125, London Wall, EC2Y 5AS.
525 B Street, Suite 1800, San Diego, CA 92101-4495, USA
225 Wyman Street, Waltham, MA 02451, USA
The Boulevard, Langford Lane, Kidlington, Oxford OX5 1GB, UK

ISBN: 978-0-12-803522-1

Library of Congress Cataloging-in-Publication Data
A catalog record for this book is available from the Library of Congress.

British Library Cataloguing-in-Publication Data
A catalogue record for this book is available from the British Library.

For Information on all Academic Press publications
visit our website at http://store.elsevier.com/

Working together
to grow libraries in
developing countries

www.elsevier.com • www.bookaid.org

CONTENTS

ACKNOWLEDGMENTS

Selected excerpts and figures from M. Levin, "Granulation: End-Point Theory, Instrumentation, and Scale-Up, Education Anytime," CD-ROM Short Course, AAPS 1999 are reprinted with permission.

Selected tables and figures from M. Levin, "Wet Granulation: End-Point Determination and Scale-Up" in *Encyclopedia of Pharmaceutical Science and Technology*, Fourth Edition, Taylor & Francis, Informa Ltd., 2013, are reprinted with permission.

About the *Expertise in Pharmaceutical Process Technology* Series

Numerous books and articles have been published on the subject of pharmaceutical process technology. While most of them cover the subject matter in depth and include detailed descriptions of the processes and associated theories and practices of operations, there seems to be a significant lack of practical guides and "how to" publications.

The *Expertise in Pharmaceutical Process Technology* series is designed to fill this void. It comprises volumes on specific subjects with case studies and practical advice on how to overcome challenges that the practitioners in various fields of pharmaceutical technology are facing.

FORMAT

- The series volumes will be published under the Elsevier Academic Press imprint in both paperback and electronic versions. Electronic versions will be full color, while print books will be published in black and white.

SUBJECT MATTER

- The series will be a collection of hands-on practical guides for practitioners with numerous case studies and step-by-step instructions for proper procedures and problem solving. Each topic will start with a brief overview of the subject matter and include an exposé, as well as practical solutions of the most common problems along with a lot of common sense (proven scientific rather than empirical practices).
- The series will try to avoid theoretical aspects of the subject matter and limit scientific/mathematical exposé (e.g., modeling, finite elements computations, academic studies, review of publications, theoretical aspects of process physics or chemistry) unless absolutely vital for understanding or justification of practical approach as advocated by the volume author. At best, it will combine both the practical ("how to") and scientific ("why") approach, based on

practically proven solid theory — model — measurements. The main focus will be to ensure that a practitioner can use the recommended step-by-step approach to improve the results of his or her daily activities.

TARGET AUDIENCE

- The primary audience includes pharmaceutical personnel, from R&D and production technicians to team leaders and department heads. Some topics will also be of interest to people working in nutraceutical and generic manufacturing companies. The series will also be useful for those in academia and regulatory agencies. Each book in the series will target a specific audience.

The *Expertise in Pharmaceutical Process Technology* series presents concise, affordable, practical volumes that are valuable to patrons of pharmaceutical libraries as well as practitioners.

Welcome to the brave new world of practical guides to pharmaceutical technology!

INTRODUCTION

Judging by the number of recently published papers on the topic of end-point determination and scale-up of wet granulation process, the subject matter is still far from being settled in the pharmaceutical industry. A lot of new ideas and technologies are being proposed, some more promising than others and some with more scientific justification.

The purpose of this book is to review some of these approaches and to justify the use of dimensional analysis as the most promising scientific way to deal with the problem. The book is the first in a planned series presenting hands-on solutions to everyday challenges facing pharmaceutical practitioners.

Understanding and implementing science-based methodology to formulation development and scale-up should be viewed as a part of quality by design (QbD) approach as defined by International Conference on Harmonization of Technical Requirements for Registration of Pharmaceuticals for Human Use (ICH). Its annex Q8 Pharmaceutical Development defines pharmaceutical QbD as a systematic approach to development that begins with predefined objectives and emphasizes product and process understanding and process control based on sound science and quality risk management. Because scale-up is an integral part of process development, it is also subject to the requirement of rigorous scientific justification based on continuous improvement and knowledge management.

It should be noted here that Scale-Up and Post-Approval Guidance (SUPAC) by the Food and Drug Administration does not give advice on the scale-up principles of development because it deals only with postapproval scale-up or scale-down. Thus, so far, there was no unifying nonempirical recommendation or approach to scale-up from small-scale to clinical batches to production in general and to scale-up of wet granulation end point in particular.

It should also be noted in this introduction that this book deals with high- or low-shear wet granulation in fixed bowl mixer-granulators. Treatment of fluid bed and other forms of granulations is outside the scope of this book, although an example of application of dimensional analysis theory to roller compaction is given to show the broad spectrum of applicability of this approach.

No special knowledge of engineering or mathematics (except for simple matrix transformations) is required from the readers of this book because any theoretical considerations are fully explained. After the general principles are laid out, the main thrust is practical applications of this scientific approach.

As a source of information on batch enlargement technique, this book will be of practical interest to formulators, process engineers, validation specialists, and quality assurance personnel, as well as production managers.

—Michael Levin, PhD
Milev, LLC Pharmaceutical Technology Consulting

CHAPTER 1

Scale-Up Basics

Scale-up is generally understood as the procedures of transferring the results of research and development obtained on a laboratory scale to the pilot plant and finally to production scale. Although many variations of the theme exist (e.g., batch size enlargement, scale down when required to improve the quality of the product, pilot scale as an intermediate step, multiplying small- or medium-scale operations to increase output), this book concentrates on classic procedures of size increase of the processing volume, specifically as applied to wet granulation.

Let's face it, most scale-up practices and procedures currently used in the pharmaceutical industry are empirical in nature.

There are always some practical "trade secrets" that are known to experienced operators, some unwritten operating procedures that are not necessarily based on any theoretical foundation. The problem with this approach is that it is empirical and thus has limited applicability to ever-changing conditions.

Numerous drawbacks and difficulties could be met on this path of scale-up. When quality of the product on the production floor is not as was envisioned in the development and preapproval stages, losses in terms of effort and money can be enormous. That is why scale-up as part of development is so important and why postapproval changes are so strictly regulated.

No unit operation is exempt from this predicament, and wet granulation is one of the most challenging processes to scale-up because of many variables involved in the process. Compared with tableting, for example, in which small- and large-scale operations differ by speed only (the same die volume and same force can be assured), in wet granulation, one has to take into account the geometric, kinematic, and dynamic differences.

How to Scale-Up a Wet Granulation End Point Scientifically. DOI: http://dx.doi.org/10.1016/B978-0-12-803522-1.00001-5

In this book, we recommend a universal approach to scale-up, a procedure called dimensional analysis. Dimensional analysis is widely used in many industries; in chemical engineering; and in applications such as fluid dynamics, heat and mass transfer, mixing of different substrates, homogenization, and so on (Zlokarnik, 2006). The method can be applied to any unit operation (in fact, any process) because any physical process can be described in terms of dimensionless variables. After this is achieved, the process becomes "scale invariable," that is, independent of scale. In other words, the key to successful scale-up is to eliminate the scale (e.g., linear dimensions, time, forces) from the process description.

For wet granulation, the scientific approach to scale-up of an acceptable end point should include (1) understanding and application of the principles of dimensional analysis, (2) instrumentation to measure critical process parameters, (3) procedures to measure and calculate dimensional numbers describing your process, (4) creating a prediction equation that will calculate the required target dimensionless Newton power number Np, and (5) implementation of a computerized monitoring and control of your process that will stop the process when the desired Np is reached.

As a side note, we should mentions that improper scale-up of granulation process sometimes can only be detected on the tableting stage. If tablets made on the production press fail dissolution tests, the granule particle size or distribution can be the culprit.

Another note relates to Food and Drug Administration (FDA) Scale-Up and Post-Approval Changes (SUPAC) guidance (FDA Guidance: Scale-up and Post-Approval Changes for Modified Release Products [SUPAC-MR], 1995 and Scale-up and Post-Approval Changes Guidance for Immediate Release Products [SUPAC-IR], 1997). Postapproval scale-up changes are subject to regulatory oversight. If the process is different after scale-up, the manufacturer has to demonstrate that the product produced by a modified process will have similar bioavailability using data such as granulation studies, finished product test results, content uniformity, and potency and dissolution profiles. Although, generally speaking, postapproval scale-up or scale-down changes and recommendations covered by the FDA SUPAC guidance can be viewed from a dimensional analysis perspective, we will limit our discussions to scale-up principles and procedures during product and process development before regulatory approval. Postapproval changes just add another layer of complexity and risk to the whole process of scale-up.

CHAPTER 2

Dimensional Analysis

Dimensional analysis is based on the assumption that a mathematical formulation of a physical process should be valid in any system of dimensions.

A dimension is a *qualitative* description of a physical entity (e.g., length or mass). A physical quantity, on the other hand, is a *quantitative* description of a physical quality (e.g., length of 10 m).

A rational approach to scale-up using dimensional analysis was devised a century ago. The procedure, based on process similarities among different scales, was first proposed by Lord Rayleigh (1915). It was first applied to pharmaceutical granulation by Leuenberger (1979).

Dimensional analysis is a method for producing dimensionless numbers that completely describe the process. Because the numbers are

How to Scale-Up a Wet Granulation End Point Scientifically. DOI: http://dx.doi.org/10.1016/B978-0-12-803522-1.00002-7

dimensionless, only their numerical value is relevant to the process state they represent; the values for a particular state should be identical at any scale. The analysis must be done before any measurements are made because dimensionless numbers essentially condense the frame in which the measurements are taken and evaluated. The dimensionless analysis method can be applied even when the equations governing the physical process are not known.

After the process is described in terms of dimensionless variables, problems of scale-up disappear because there is no scale; the process is characterized solely by the numerical values of the dimensionless numbers.

Small-scale process could be viewed as a model of the larger scale process.

2.1 THEORY OF MODELING AND PRINCIPLE OF SIMILITUDE

Scale-up from a small to larger scale can be viewed as a modeling process. Small scale is "a model" of the larger scale process. According to theory of modeling, two or more processes may be considered completely similar if there is a geometrical, kinematic, and dynamic similarity ("principle of similitude" of Buckingham, 1914). Under the condition of similarity, all the dimensionless numbers required to describe the process at any given state will have the same numerical value (Zlokarnik, 1991).

2.1.1 Geometrical Similarity

Two systems are called geometrically similar if they have the same ratio of characteristic linear dimensions. For example, two cylindrical mixing vessels are geometrically similar if they have the same ratio of height to diameter.

2.1.2 Kinematic Similarity

Two geometrically similar systems are called kinematically similar if they have the same ratio of velocities between the corresponding system points. For example, the ratio of impeller tip speed to that of the impeller rotational speed is usually fixed by virtue of a fixed blade diameter.

2.1.3 Dynamic Similarity

Two kinematically similar systems are dynamically similar when they have the same ratio of forces between the corresponding points. Dynamic similitude for wet granulation would imply that the wet mass flow pattern is in the bowl are similar.

Often the problem with application of dimensional analysis lies in the fact that the scale-up effort is applied to processes that are not geometrically similar.

For example, Collette Gral 10, 75, and 300 mixers are not geometrically similar (Faure et al., 1999). In such cases, a proper correction to the resulting equations is required. Alternatively, one can subdivide the process into several parts, each of which can be treated under the conditions of similarity.

2.2 DIMENSIONLESS NUMBERS

Dimensionless numbers most commonly used to describe the wet granulation process are Newton (Eq. 2.1), Froude (Eq. 2.2), and Reynolds (Eq. 2.3):

$$N\mathrm{p} = P/(\rho \times n^3 \times d^5) \tag{2.1}$$

$$Fr = n^2 \times d/g \tag{2.2}$$

$$Re = d^2 \times n \times \rho/\eta \tag{2.3}$$

(For list of symbols, notation, and dimensions, see Chapter 6.)

The Newton (power) number, which relates the drag force acting on a unit area of the impeller and the inertial stress, quantifies power required to overcome friction in fluid flow in a stirred vessel. In mixer-granulation applications, this number can be calculated from the torque or power consumption of the impeller or estimated from the power consumption of the motor.

Froude number, first introduced by Merrifield (1870), was described as interplay of the centrifugal force (pushing the particles against the mixer wall) and the centripetal force produced by the wall, creating a "compaction zone."

Reynolds number relates the inertial force to the viscous force. The number bears the name of Osborne Reynolds who popularized its use (Reynolds, 1883). As such, it can be used to determine dynamic similitude between different scales of viscous flow.

2.3 Π-THEOREM

The so-called Π-theorem, or Buckingham's theorem (Buckingham, 1914) states that every physical relationship between n dimensional variables and constants can be reduced to a relationship $f(\Pi_0, \Pi_1, .s . ., \Pi_m) = 0$ between $m = n - r$ mutually independent dimensionless groups (numbers), where r is the number of dimensional units, i.e., fundamental units (rank of the dimensional matrix). In other words, if you have a set of n dimensional variables describing a process, this set can be transformed into a set of m dimensionless variables describing the same process, with m being less than n by the number of dimensions pertaining to the process. The theorem essentially states that the number of parameters required to define the problem can be reduced by the number of dimension if the parameters are expressed in dimensionless form. The meaning of this theorem will become clear after seeing examples of its applications below.

2.4 RELEVANCE LIST

A list of all physical parameters and variables thought to be important for the process is the starting point of the dimensional analysis ("'relevance list"). To set up a relevance list for any process, one needs to compile a *complete* set of all relevant and mutually independent variables and constants that influence the process.

All entries in the list can be subdivided into geometric, physical, and operational. Each relevance list should include one target (dependent "response") variable.

2.5 DIMENSIONAL MATRIX

After the relevance list is created, dimensional analysis can be simplified by arranging all variables from the relevance list in a matrix form, with a subsequent transformation yielding the required dimensionless numbers. The dimensional matrix consists of a square core matrix and a residual matrix. (You will see how it is done in the examples below.)

The rows of the matrix represent of the basic dimensions, such as length (L), mass (M), and time (T). The columns consist of the physical quantities from the relevance list. The columns of the core matrix are holding critical process variables, and the number of columns in the core matrix is equal to the number of basic dimensions, so that the core matrix is always a square one. The most important physical properties and process-related parameters, as well as the "target" variable (i.e., the one we would like to predict on the basis of other variables) are placed in the columns of the residual matrix.

The core matrix is then linearly transformed into a matrix of unity in which the main diagonal consists only of ones and the remaining elements are all zero. The dimensionless numbers are generated by dividing the variables in the residual matrix by the variables of the unity matrix with the exponents indicated by the unit values of residual matrix. What sounds somewhat complicated is, in fact, a rather simple process, as will be illustrated below in the examples.

2.6 EXAMPLES

2.6.1 Wet Granulation Example

We take this example from a paper by Landin et al. (1996). Wet granulation scale-up has been studied by Ray Rowe and Michael Cliff's group using the dimensionless numbers of Newton (power), Reynolds, and Froude. The analysis led to pinpointing of end point in geometrically similar high-shear Fielder PMA 25-, 100-, and 600-L mixers. The relevance list for this process is given in Table 2.1.

The list includes power consumption of the impeller (as a response variable) and six other quantities, namely, impeller diameter, impeller speed, vessel height, specific density and dynamic viscosity of the wet mass, and the gravitational constant.

The inclusion of a universal physical constant, such as the acceleration due to gravity (gravitational constant), in the relevance list is a must if it has an impact on the process (e.g., wet granulation process on the moon will be different from the one here on Earth, all other variables being equal).

Table 2.1 Relevance List for a Typical Wet Granulation Process				
	Quantity	Symbol	Units	Dimensions
1	Power consumption	P	Watt	$M\ L^2\ T^{-3}$
2	Specific density	ρ	kg/m^3	$M\ L^{-3}$
3	Blade diameter	D	m	L
4	Blade speed	n	rev/s	T^{-1}
5	Dynamic viscosity	η	Pa*s	$M\ L^{-1}\ T^{-1}$
6	Gravitational constant	g	m/s^2	$L\ T^{-2}$
7	Bowl height	H	m	L

Table 2.2 Dimensional Matrix for a Typical Wet Granulation Process							
	Core Matrix			Residual Matrix			
	ρ	d	n	P	η	g	H
Mass M	1	0	0	1	1	0	0
Length L	−3	1	0	2	−1	1	1
Time T	0	0	−1	−3	−1	−2	0

Note the line item of dynamic viscosity, making it applicable to viscous binders and allowing long-range particle interactions responsible for friction.

From the variables in the relevance list, a dimensional matrix (Table 2.2) is constructed.

The dimensional matrix was constructed by the rows listing the basic dimensions and the columns indicating the physical quantities from the relevance list.

By a simple linear transformation, this matrix is converted to a unity matrix and a resulting residual matrix (Table 2.3).

There is a total of 7 variables, 3 basic dimensions (M, L, T), and 4 columns in the residual matrix; therefore, four dimensionless Π groups (numbers) will be formed, in accordance with the Buckingham's Π-theorem (7 variables − 3 dimensions = 4 dimensionless groups). The groups (dimensionless variables) are obtained by from the dividing each element of the residual matrix by the column headers of the unity

	Core Matrix			Residual Matrix			
	ρ	d	n	P	η	g	H
M	1	0	0	1	1	0	0
3M + L	0	1	0	5	2	1	1
−T	0	0	1	3	1	2	0

Table 2.3 Unity and Residual Matrices for a Typical Wet Granulation Process

matrix, with the exponents indicated in the residual matrix, as follows (Eqs. 2.4 to 2.7):

$$\Pi_0 = P/(\rho^1 \cdot d^5 \cdot n^3) = N_p \quad \text{Newton (power) number} \qquad (2.4)$$

$$\Pi_1 = \eta/(\rho^1 \cdot d^2 \cdot n^1) = Re^{-1} \quad \text{Reynolds number} \qquad (2.5)$$

$$\Pi_2 = g/(\rho^0 \cdot d^1 \cdot n^2) = Fr^{-1} \quad \text{Froude number} \qquad (2.6)$$

$$\Pi_3 = H/(\rho^0 \cdot d^1 \cdot n^0) = H/d$$
$$\text{Geometric number (ratio of characteristic lengths)} \qquad (2.7)$$

From the theory of modeling, we know that the above dimensional groups are functionally related. The form of this functional relationship f, however, can be established only through experiments.

We know from the dimensional analysis theory that under the assumed conditions of dynamic similarity, one (target) group can be expressed as a function of all others, that is:

$$\Pi_0 = f\left(\Pi_1, \Pi_2, \Pi_3\right) \text{ and, therefore, } N_p = f(Re, Fr, H/d).$$

When the conditions of dynamic similarity are not assured, some corrections should be made. Thus, when Landin et al. (1996) applied corrections for gross vortexing, geometric dissimilarities, and powder bed height variation, data from Fielder PMA 25-L mixer allowed predictions of optimum end-point conditions at 100- and 600-L mixers.

A linear regression of Newton number (power) on the product of the Reynolds number, Froude number, and geometric number (in log/log domain) yields an equation of the form (Eq. 2.8):

$$\text{Log}_{10} N_p = a \cdot \log_{10}(Re \cdot Fr \cdot H/d) + b \qquad (2.8)$$

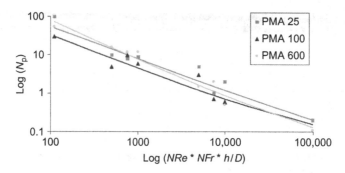

Figure 2.1 Regression lines of the Newton power number on the product of the Reynolds number, Froude number, and the length ratio for three different Fielder mixers. (© 1996 Elsevier reprinted with permission.)

Tests performed on a single small scale are sufficient to determine the functional relationship that will apply to any scale.

The (dimensionless) regression line represented by this equation (Fig. 2.1) contains all possible states of the process, including the end points, however defined, and therefore can be used for end-point determination, reproduction, and scale-up.

After an end point is reached, it is characterized by certain set of numerical values of the dimensionless variables, and these values will be independent of scale because the variables are dimensionless. At the same end point, no matter how defined, the rheological and dimensional properties of the granules are similar. This means that the density and dynamic viscosity of the wet mass are fixed for each end point, and the only variables that are left to vary are the process variables, namely batch mass, impeller diameter and speed, and the geometry of the vessel. In other words, how we reached the end point is not critical; the important thing is that we have arrived at our destination.

2.6.2 Dry Granulation (Roller Compaction) Example

A simple relevance table for a unit roller compaction operation is presented in Table 2.4.

Here we have two material property variables (true density and compressibility factor), three process parameters (roll speed, screw speed, and roll diameter), two resulting response variables (gap and normal stress), and one target variable (ribbon density).

	Quantity	Symbol	Units	Dimensions
	Table 2.4 Relevance Table for a Typical Dry Granulation Process			
1	True density of material	ρ	kg/m³	$M\,L^{-3}$
2	Roll speed	R	rev/min	T^{-1}
3	Gap	G	m	L
4	Ribbon density	d	kg/m³	$M\,L^{-3}$
5	Normal stress on the roll	P	MPa	$M\,L^{-1}\,T^{-2}$
6	Roll diameter	D	m	L
7	Compressibility of material	k	1/MPa	$M^{-1}\,L\,T^{2}$
8	Screw speed	S	Rev/min	T^{-1}

Table 2.5 Dimensional Matrix for a Typical Dry Granulation Process

	Core Matrix			Residual Matrix				
	ρ	R	G	d	P	D	k	S
M	1	0	0	1	1	0	−1	0
L	−3	0	1	−3	−1	1	1	0
T	0	−1	0	0	−2	0	2	−1

Normal stress on the roll can be measured by a special sensor and expressed as force exerted on the ribbon per unit area. Adding roll diameter to the list of relevant factors allows one to account for the influence of the contact time in the nip area. This makes the resulting equation applicable to roller compactors of different size.

Compressibility factor represents a relative change of the powder volume with pressure (Eq. 2.9):

$$k = \Delta V / (V \cdot \Delta P) \qquad (2.9)$$

where $\Delta V/V$ is a relative volume change in response to change in pressure ΔP.

Adding compressibility factor to the list of relevant factors allows one to account for the influence of the powder properties (in addition to the true density). This makes the resulting equation applicable to powders with different compressibility properties.

Dimensional matrix (Table 2.5) is constructed in a manner similar to that of the previous example.

Table 2.6 Unity and Residual Matrices for a Typical Dry Granulation Process

	Unity Matrix			Residual Matrix				
	ρ	R	G	d	P	D	k	S
M	1	0	0	1	1	0	−1	0
−T	0	1	0	0	2	0	−2	1
L + 3M	0	0	1	0	2	1	−2	0

Again, a simple linear transformation yields a unity and residual matrices of Table 2.6.

Thus, according to the Π-theorem, we get 5 dimensionless groups (8 variables minus 3 basic dimensions, Eqs. 2.10 to 2.14):

$$\Pi_0 = d/\rho \quad \text{Solid fraction} \tag{2.10}$$

$$\Pi_1 = P/(\rho^1 \cdot R^2 \cdot G^2) \tag{2.11}$$

$$\Pi_2 = D/G \tag{2.12}$$

$$\Pi_3 = k/(\rho^{-1}\cdot R^{-2}\cdot G^{-2}) = k\cdot\rho\cdot R^2\cdot G^2 \tag{2.13}$$

$$\Pi_4 = S/R \tag{2.14}$$

The dimensionless group Π_0, containing the target quantity, can be expressed as a function of all other groups (Eq. 2.15):

$$\Pi_0 = f\left(\Pi_1, \Pi_2, \Pi_3, \Pi_4\right) \tag{2.15}$$

The actual functional relationship can be specified from data, for example, data may be fitted to Eq. 2.16:

$$d/\rho = A\cdot\left(\Pi_1\cdot\Pi_2\cdot\Pi_3\cdot\Pi_4\right)^B \tag{2.16}$$

where 'A' and 'B' are regression coefficients. Regression will be linear in a log/log domain. The functional relationship can be simplified as (Eq. 2.17)

$$d/\rho = f(k/P, G/D, S/R) \tag{2.17}$$

2.6.3 Compaction (Tableting) Example
Here is an example adapted from Levin and Zlokarnik (2011). The relevance list for a tableting process is presented in Table 2.7.

	Quantity	Symbol	Units	Dimensions
	Table 2.7 Relevance Table for a Typical Tableting Process			
1	Tablet tensile strength	H	MPa	$M\,L^{-1}\,T^{-2}$
2	Die diameter	d	m	L
3	Depth of fill	h	m	L
4	Dwell time	τ	s	T
5	Maximal applied pressure	P	MPa	$M\,L^{-1}\,T^{-2}$
6	Compressibility of material	k	1/MPa	$M^{-1}\,L\,T^{2}$
7	Press speed	n	rev / s	T^{-1}

The table includes one target variable (tensile strength, or hardness), two geometric parameters (die diameter—assuming cylindrical tablet shape—and depth of fill), one material property (compressibility factor), and three process-related variables (applied pressure, dwell time, and press speed).

Compressibility factor is a lumped parameter, combining the influence of a number of powder properties, such as flowability, moisture content, particle size distribution, morphology, and so on.

The dwell time is defined as the time when the flat portion of punch head is in contact with the compression roll. Thus, for each tooling type, it is a derivative of the rotational speed of the turret and is a characteristic of the tooling rather than press speed.

From Table 2.7, construction of a dimensional matrix (Table 2.8) is straightforward.

A simple linear transformation yields the unity and residual matrices (Table 2.9).

Application of the Π-theorem leads to 4 dimensionless groups (7 dimensional variables − 3 dimensions, Eqs. 2.18 to 2.21):

$$\Pi_0 = H \cdot k \qquad (2.18)$$

$$\Pi_1 = P \cdot k \qquad (2.19)$$

$$\Pi_2 = \tau \cdot n \qquad (2.20)$$

$$\Pi_3 = h/d \qquad (2.21)$$

Table 2.8 Dimensional Matrix for a Typical Tableting Process						
	Core Matrix			Residual Matrix		
	k	n	d	H	P	τ	h
M	−1	0	0	1	1	0	0
L	1	0	1	−1	−1	0	1
T	2	−1	0	−2	−2	1	0

Table 2.9 Unity and Residual Matrices for a Typical Tableting Process						
	Unity Matrix			Residual Matrix		
	κ	n	d	H	P	τ	h
−M	1	0	0	−1	−1	0	0
−T−2M	0	1	0	0	0	−1	0
L + M	0	0	1	0	0	0	1

Thus the dimensional analysis of a tableting process yields the following relationship (Eq. 2.22):

$$\Pi_0 = f\left(\Pi_1, \Pi_2, \Pi_3\right) \tag{2.22}$$

The actual functional relationship can be specified from data, for example, data may be fitted to the Eq. 2.23:

$$H \cdot k = a \cdot (P \cdot k \cdot \tau \cdot n \cdot h/d)^{b} \tag{2.23}$$

where 'a' and 'b' are regression coefficients. Regression will be linear in a log/log domain.

Alternatively, data may fit Eq. 2.24 of the form

$$H \cdot k(\tau \cdot n)^{a} = f(P \cdot k, h/d) \tag{2.24}$$

The type of powder determines the influence of the dimensional process number $n \cdot \tau$ of the tablet press.

2.7 COMPARISON OF ATTAINABLE FROUDE NUMBERS

For the same end point, in dynamically similar mixers (same geometrical ratios and same flow patterns), all dimensionless numbers describing the system should have the same numerical value. That means that

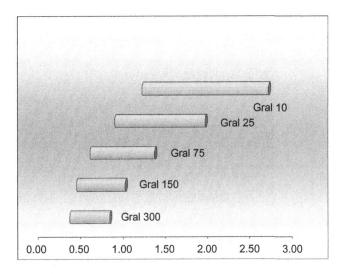

Figure 2.2 Ranges of possible Froude numbers for Collette-Gral mixers.

Newton, Froude, and Reynolds numbers should keep the same value for the small- and large-scale processes.

There is sufficient evidence in literature indicating that an end point in mixers-granulators could be reproduced and scaled up by keeping the Froude numbers constant.

Because each mixer has a range of attainable Froude numbers, moving the end point from one scale to another between mixers can only be achieved without problems when such ranges overlap. These considerations can be useful for planning a scale-up or technology transfer operation.

Figure 2.2 represents such a range of all possible Froude numbers for Collette-Gral mixers. From this comparison, it is evident that that Gral 10 and Gral 150 have no overlap of Froude number ranges; therefore, a direct scale-up is not possible (in addition, Gral mixers are not exactly similar geometrically, as stated earlier).

From the range of attainable Froude numbers for Fielder PMA series mixers (Fig. 2.3), one can see that the 10-L laboratory scale mixer at its lowest speed settings can reach the Froude numbers of PMA 300, so that any scale-up effort beyond that scale should be done in stages.

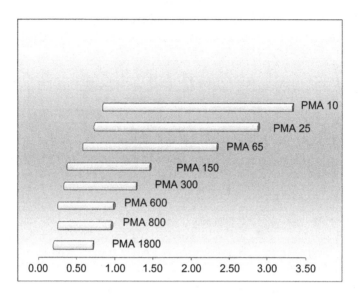

Figure 2.3 Ranges of possible Froude numbers for Fielder PMA mixers.

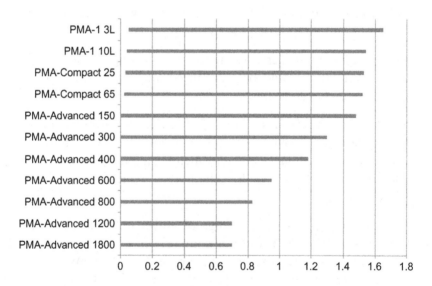

Figure 2.4 Ranges of possible Froude numbers for new generation of Fielder PMA MIXERS.

Newer Aeromatic-Fielder mixers have variable speed drives, allowing the user greater flexibility in terms of operating speed and power input. The ranges of Froude numbers for these mixers (Fig. 2.4) look more uniform from the scale-up point of view.

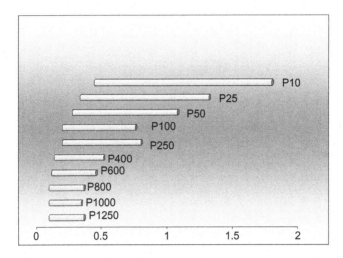

Figure 2.5 Ranges of possible Froude numbers for Diosna mixers.

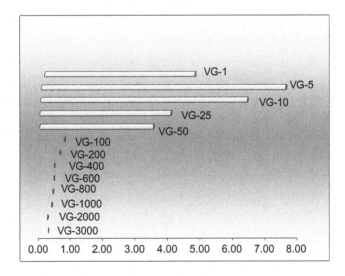

Figure 2.6 Ranges of possible Froude numbers for Powrex mixers.

The situation with Diosna mixers is somewhat familiar (Fig. 2.5): one can scale-up from 10 L laboratory scale to 250 L as an intermediate step to reach the same end point at 400 L and beyond.

Froude number ranges for Powrex mixers (Fig. 2.6) show a better alignment with the smallest 1 L vessel range including practically all Froude numbers for larger volumes. However, it should be noticed that the volumes of 100 L and above have fixed speed.

CHAPTER 3

Wet Granulation

How to Scale-Up a Wet Granulation End Point Scientifically. DOI: http://dx.doi.org/10.1016/B978-0-12-803522-1.00003-9

3.3 END-POINT DETERMINATION AND SCALE-UP: SEMI-EMPIRICAL METHODS

3.3.1 Relative Swept Volume
3.3.2 Impeller Tip Speed
3.3.3 Experimental Design
3.3.4 Examples

3.1 BASICS

Wet granulation is used to improve the flow, compressibility, bioavailability, and homogeneity of low-dose blends, electrostatic properties of powders and the stability of dosage forms. It also helps to prevent segregation of the blend components. Compared with a low-shear or fluid bed process, high-shear granulation requires more fluid binder, resulting in longer drying time but also in a more cohesive material (Parikh, 2010).

Main factors in wet granulation include the densification, agglomeration, shearing, and compressing action of the impeller. Because of rapid densification and agglomeration caused by the shearing and compressing action of the impeller in a high-shear mixer, granulation and wet massing can be done relatively quickly. However, there is always a possibility of overgranulation caused by excessive wetting. This would lead to production of low-porosity granules, which in turn affects the mechanical properties of the tablets.

Numerous theoretical models have been proposed to describe the high-shear wet granulation process (Kumar et al., 2013). However, both population balance modeling and discrete element modeling are still presenting open questions about the gap in our understanding of micro- and macro-scales of the process.

Three major stages of the process are mixing, granulation, and wet massing. The three key areas of wet granulation processes are wetting and nucleation, consolidation and growth, and breakage and attrition.

3.1.1 Granulation Stage

During granulation, liquid bridges between particles are formed. This process is followed by particle coalescence and subsequent breakage of the bonds. Particles change specific surface area and moisture content, as well as mean granule size and apparent viscosity.

3.1.2 Wet Massing Stage

During the wet massing stage, the intragranular porosity decreases. At the same time, granules usually increase in size. Liquid saturation affects intragranular porosity, and heat generated during the process causes evaporation, leading to subsequent decrease in the mean granule size, especially in small-scale mixers. Wet massing time is a key process parameter in high-shear granulation because it can simultaneously improve powder flowability and decrease granule tabletability (Shi et al., 2011).

3.1.3 Process of Agglomeration

It was suggested (Nichols et al., 2002) to use term the "agglomeration" to describe an assemblage of particles in a powder and that the term "aggregate" should be confined to prenucleation structures. Agglomeration of particles in wet granulation has been studied extensively (Kristensen, 1996, Knight et al., 1998).

Forces involved in wet granulation can be identified as acceleration F1, frictional F2, centripetal F3, and centrifugal F4 (Fig. 3.1). The interplay of these forces creates the dynamic agglomeration pattern of each granulation batch and is responsible for the resulting granule apparent viscosity and wet mass consistency. Mix consistency and reproducibility are controlled by the end point and the load on the main impeller at that end point.

Binder addition rate controls granule density, impeller and chopper speed control, granule size, granulation rate, and dry granule strength (Cliff, 1990). Other factors that affect the granule quality include spray position and spray nozzle type and, of course, the product composition. Quantity and feeding rate of granulating liquid determine the particle size of granulate. Such variables as mixing time and bowl or

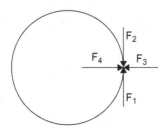

Figure 3.1 Forces involved in wet granulation.

product temperature are not independent factors of the process but rather are responses of the primary factors listed.

It stands to reason that mean granule size is strongly dependent on the specific surface area of the excipients, as well as the moisture content and liquid saturation of the agglomerate.

Granule tabletability deteriorates significantly when initial moisture content of powder increases while all other processing parameters remain unchanged (Shi et al., 2011). The reduced tabletability is largely caused by increased granule size.

Chitu et al. (2011) showed that optimum liquid requirement for granulation varies with binder type and decrease with increasing viscosity, and granule growth kinetics has been found to be to be related to the work of adhesion for low-viscosity binders. Granule strength has been evaluated for wet granules by the means of wet mass rheometry. Consistency measurements using mixer torque rheometer and tests of mechanical strength of dried granules by means of uniaxial compression showed that for low-viscosity binders, both wet mass consistency and dry granule strength depend on the work of adhesion. For high-viscosity binders, higher wet mass consistencies but lower dry granule strengths have been observed.

The method of binder addition can play a critical role. The questions are (1) Should dry binder be mixed with the powder with the subsequent addition of water, or it is better to add binder solution? (2) Should liquid be added slowly and continuously or all at once? (3) What method of binder addition is better: pouring, pumping, or atomized spray? And (4) Does nozzle shape have any effect on the granulation process?

There is no consensus on the best way to add binder. Holm (1987) recommends adding dry binder to fluid rather than to the mix to assure homogeneity of binder distribution. Others recommend just the opposite (e.g., Laicher et al., 1997).

Slow continuous addition of water (in case the water-soluble binder is dry mixed) or a binder solution to the mix is a granulation method of choice (Leuenberger et al., 1979). The rate of liquid binder addition can be controlled by a peristaltic pump. It is recommended to keep the rate at a minimum to avoid local overwetting (Werani, 1988).

Horsthuis et al. (1993) recommended adding binder liquid all at once to ensure the ease of processing and reproducibility, to reduce processing time, and to avoid wet mass densification that may occur during the liquid addition. This latter phenomenon may obscure the scale-up effect of any parameter under investigation.

The method of addition (pneumatic or binary nozzle, atomization by pressure nozzle) should be preserved in any end-point scale-up effort.

The means of adding binder will have differing degrees of effects on particle size and shape depending on the type of formulation used for high-shear granulations.

3.2 WHAT CAN BE MEASURED ON A MIXER-GRANULATOR?

Generally speaking, the benefits of mixer instrumentation extend well beyond a possible use for end-point determination and scale-up. Similar to instrumentation of any other equipment, sensors installed on a mixer-granulator can be used to perform machine troubleshooting (e.g., detect worn-out gears and pulleys), identify mixing and binder irregularities, make formulation fingerprints (when a monitored and logged batch record becomes a batch and mix ID), and help in raw material evaluation and general process optimization.

In what follows, we will briefly identify some of the ways to monitor the wet granulation process.

3.2.1 Current

For small-scale mixers with direct current (DC) motors, load on the main impeller may be proportional to current in some intervals (Cliff, 1990); therefore, a current meter (ammeter) can be used to monitor a wet granulation process. However, for alternating current (AC) motors (most often used in modern and larger mixers), there may be no significant change in the current because motor load varies up to 50% of full scale. At larger loads, the current draw may increase, but this increase is not linearly related to load, and therefore, monitoring current will not give you accurate measurement of load. Sochon et al. (2010) monitored current draw of the impeller motor on a Fielder PMA 65-L mixer-granulator and reported that this measurement did not explain the resulting variation in the end product.

3.2.2 Voltage

Voltage measurement generally has no relation to load on the impeller.

3.2.3 Capacitance

Physically, capacitance is a measure of the capacity to store electric charge for a given voltage potential difference. Capacitive sensor can generate data related to the moisture distribution and granule formation (Corvari et al., 1992). It can be threaded into an existing thermocouple port for in-process monitoring and provide similar end points under varying rates of agitation and liquid addition.

3.2.4 Conductivity

Conductivity of the damp mass (Fry et al., 1987) can be used to quantify the uniformity of liquid distribution and packing density during wet massing time.

3.2.5 Chopper Speed

A chopper is usually needed for lower impeller speeds to do what its name suggests: to chop larger chunks of material to facilitate wetting and agglomeration. Chopper speed has no significant effect on the mean granule size (Holm, 1987).

3.2.6 Motor Slip and Motor Load Analyzer

Motor slip is the difference between the rotational speed of an idle motor and the motor under load (Timko et al., 1987). As such, motor slip measurements, although relatively inexpensive, do not offer advantages over the power consumption measurements. The method did not gain popularity, probably because the slip is not always linearly related to the load (Fink and Beaty, 1993), despite some claims to the contrary.

Of all these early attempts to monitor granulation, most have only historical value. Motor current and or voltage are still used by some original equipment manufacturers (OEMs) as an indicator of the load on the main motor, but as will be shown later, these measurements have little predictive value.

3.2.7 Temperature

Product and jacket temperature are usually measured by thermocouples. These response variables are controlled by a variety of factors, notably, the speed of the main impeller and the rate of the binder addition.

The ultimate goal of any measurement in a granulation process is to estimate the viscosity and density of the granules and perhaps to obtain an indication of the particle size mean and distribution. An indirect method to get estimates of these granulate properties is to monitor load on the main impeller.

The most popular measurements that can monitor the load on the main impeller are direct and reaction torque and the power consumption of the main motor. In the following section, we will examine the benefits and relative disadvantages of the torque and power measurements.

3.2.8 Power and Torque
3.2.8.1 Power Consumption
Power consumption of the main mixer motor is one of the most popular and economical measurements. A wattmeter or a power cell can be connected to the motor to measure transversive voltage between two radial sides of a current conductor in a magnetic field, an effect discovered by E.H. Hall in 1879.

The measurement gained popularity since the early work of Hans Leuenberger (Leuenberger, 1979) and subsequent work, Holm (1983) and subsequent work, Landin et al. (1996), Laicher (1996), Faure et al. (1999a), Betz et al. (2004), and many others because the measurement is not expensive, does not require extensive mixer modifications, and is well correlated with the granule growth.

Power consumption and intragranular porosity values correlate with the mean granule size (Holm et al., 1985b), although the correlation is not always linear in the entire range. Power profile integrated over time may relate to the normalized work of granulation, which can accurately determine end points and is correlated well with properties of granulates (Ritala et al., 1988). Power consumption and temperature measurements can be used (Belohlav et al., 2007) for Active Pharmaceutical Ingredient (API) particle size and shape detection and for evaluation of raw material suppliers.

One has to keep in mind that power consumption measurement reflects load on the motor rather than on the impeller. It relates to the overall mixer performance, depends on the motor efficiency, and can change with time regardless of the load. Motor power consumption is nonlinearly related to the power transmitted to the shaft (Holm, 1997).

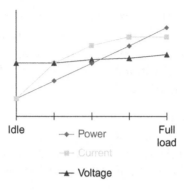

Figure 3.2 Watt-meters versus ammeters.

and the degree of this nonlinearity depends on overall mechanical condition of the mixer.

Motor power consumption is a product of current, voltage, and the so-called power factor (Fig. 3.2). In the range of interest, motor power consumption is proportional to load on the motor. Moreover, generally speaking, it does present some measure of the load on the impeller.

It is common knowledge in electrical engineering that up to 30% of the power consumption of a motor can be attributed to no-load losses because of windage (by cooling fan and air drag), friction in the bearings, and core losses that comprise hysteresis and eddy current losses in the motor magnetic circuit. Load losses include stator and rotor losses (resistance of materials used in the stator, rotor bars, magnetic steel circuit) and stray load losses such as current losses in the windings (Hirzel, 1992).

In wet granulation applications, it is advisable to use net power consumption change, that is, the absolute value of power minus the baseline no-load (empty bowl, or dry mix) reading. Nevertheless, even a baseline signal may be confounded by a possible nonlinearity of friction losses with respect to the load (Elliott, 1993). Even though the current draw of the motor increases nonlinearly with load, it does increase and generates heat in the process, further impacting the power consumption. The baseline power consumption may shift if an empty mixer is run for several hours. Moreover, because the motor efficiency drops with age, the baseline most definitely shifts over time.

3.2.8.2 Impeller Torque

In the mixing process, changes in impeller torque and power requirements occur as a result of an increase of cohesive force or the tensile strength of the agglomerates in the moistened powder bed. There are several engineering approaches to torque measurements.

3.2.8.2.1 Direct Torque Transducer

Direct torque measurements require machining of the impeller shaft or the coupling between the motor and impeller shaft to accommodate bonding of strain gages. Because the shaft is rotating, a device called slip ring is used to transmit the signal to the stationary data acquisition system (Fig. 3.3). A slip ring is composed of two concentric rings, an inner one rotating while the outer ring is stationary. Signal and excitation voltage are transmitted to data acquisition hardware via silver brushes between the rings.

Impeller torque is an excellent in-line measure of the load on the main impeller (Cavinato et al., 2010a), a fine example of what today is called process analytical technology (PAT), providing real-time data for process understanding, analysis and control (Food and Drug Administration PAT Guidance, 2004).

Impeller torque correlates well with wet particle size distribution but poorly with dry particle size because of the inherent variability of the drying process (Mackaplow et al., 2000). This is important for evaluating the results of torque monitoring for end-point determination.

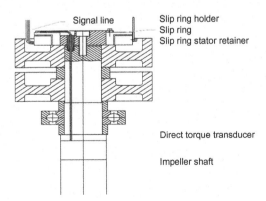

Figure 3.3 Direct torque transducer.

3.2.8.2.2 Torque Rheometer

This is an offline technique of measuring rheological properties of the granulation (Hancock et al., 1991; Landin et al., 1995; Hariharan et al., 2002; Sakr et al., 2012). The torque values obtained from rheometer were shown to be proportional to a kinematic (rather than dynamic) viscosity (Rowe and Parker, 1994). Therefore, unlike impeller torque, rheometer torque results can have a limited use in the dimensional analysis of the process. Use of such results for calculation of Reynolds numbers renders the latter to become dimensional. Torque rheometer readings were termed "measure of wet mass consistency" (Faure, 1998, 1999b; Parker et al., 1990), and the Reynolds numbers calculated from torque rheometer data is referred to as the "pseudo-Reynolds" dimensional number.

3.2.8.2.3 Reaction Torque Transducer

As the impeller shaft rotates, the motor tries to rotate in the opposite direction because, by the third law of Newton, for every force, there is a counterforce, collinear, equal and opposite in direction. But the motor is bolted in place. The tensions in the stationary motor base can be measured by a reaction torque transducer (Fig. 3.4).

Reaction torque transducer is easier to install than a direct torque sensor. It is recommended for mixers that have the motor and impeller shafts axially aligned (in these cases, the signal is equal to direct torque and opposite in sign).

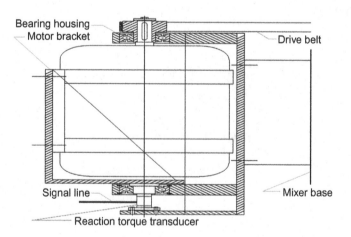

Figure 3.4 Reaction torque transducer.

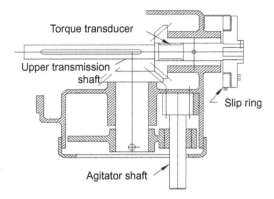

Figure 3.5 Torque transducer for planetary mixer.

3.2.8.2.4 Planetary Mixer Instrumentation
Planetary mixer instrumentation for direct torque measurement should take into account the planetary motion in addition to shaft rotation (Fig. 3.5).

The motor, via a variable speed gearbox, rotates the upper transmission shaft. The first miter gear, mounted on the upper transmission shaft, transmits the rotation to the second miter gear. This gear rotates the planetary gear carrier around a vertical axis that coincides with the axis of the bowl.

A planetary gear is installed on the carrier eccentrically to the bowl axis, and together with the agitator shaft, it simultaneously orbits around the axis of the bowl and spins around its own axis. The torque transducer, mounted on the upper transmission shaft, measures the total torque of both rotations. Similarly, a power transducer would measure the power consumed by the motor for both rotations.

3.2.8.3 Torque Versus Power
"Power consumption" usually refers to the power consumption of main motor. It represents useful work as well as the power needed to run the motor itself (e.g., losses because of eddy currents, friction in couplings).

The rotating impeller also consumes power. Obviously, the impeller requires less power than the motor. Impeller power consumption can be calculated as a product of the direct torque, rotational impeller speed, and a coefficient (usually equal to 2π times a unit conversion factor, if required):

$$\text{Power} \sim \text{Torque} \times \text{Speed} \qquad (3.1)$$

Chitu et al. (2011) showed that torque curves recorded during granulation allow good control of the process.

Figure 3.6 illustrates a typical power and torque profiles that start with a dry mixing stage, rise steeply when binder solution is added, level off into a plateau, and finally exhibit an overgranulation stage.

The power and torque signals are correlated and have similar shapes. The pattern shows a plateau region at which power consumption or torque is relatively stable.

Because the signals start with a flat line, then goes up and then flattens again, the rate of signal change (signal derivative with respect to time) must start from zero, go up, and then return to zero, meaning that there is a maximum in between.

The peak of the derivative indicates the inflection point of the signal (where the signal changes its concavity). According to granulation theory proposed by Leuenberger (1979 and subsequent work), usable granulates can be obtained in the region that begins with the peak of the signal derivative and extends into the signal plateau area. Before the inflection point, the process may require variable quantities of liquid to reach any defined state. This batch-to-batch variability of

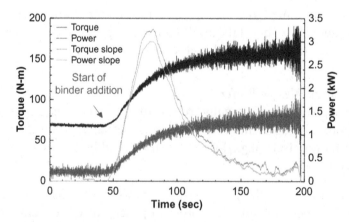

Figure 3.6 Torque and power consumption signals.

granulating liquid requirement may come from variability of raw materials, powder particle size distribution, and so on. After the inflection point (peak of derivative), the process is well defined, and the amount of binder solution required to reach a desired end point may be more or less constant. This means that at a fixed rate of binder addition, the process can be timed and stopped at a predetermined elapsed time.

The energy of mixing can be estimated by the area under the torque-time curve and can be used as an end-point parameter. The area under power consumption curve divided by the load gives the specific energy consumed by the granulation process. This quantity is well correlated with the relative swept volume (Schaeffer et al., 1986). It is also correlated with the temperature rise during mixing and wet massing.

Fluctuations of torque and power consumption are influenced by granule properties (particle size distribution (PSD), shape index, and apparent density) and the granulation time. The distribution and intensity of spectrum obtained by fast Fourier transform (FFT) analysis can be used for end-point determination (Watano et al., 1995). FFT representation of high-frequency oscillations caused by direct impact of particles on the impeller is shown in Figure 3.7.

Terashita et al. (1990) observed that frequency distribution of a power consumption signal in the end-point region reaches a steady

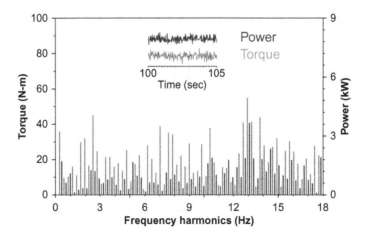

Figure 3.7 Frequency distribution of power consumption and torque signals in a wet granulation process.

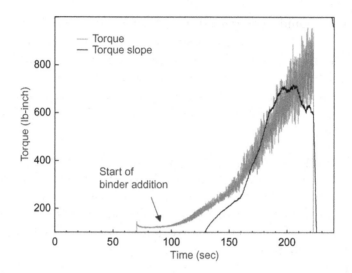

Figure 3.8 Typical granulation batch by an experienced operator (torque profile).

state. Under the conditions of this steady state, the number of breaking agglomerates approximately equals the number of forming ones to yield a time-independent final-size distribution. Experimental evidence to support the theoretical findings is obtained during the present research by measuring particle size distributions at line at crucial points during granulation of a typical pharmaceutical powder in a high-shear mixer. To reach a steady state, binder addition has to be slow enough, and wet massing has to be long enough so that neither has an influence on the final properties of the granules.

A number of publications describe practical experience of operators on the production floor (Record, 1979; Werani, 1988). Figures 3.8 to 3.11 illustrate the fact that monitoring torque or power can fingerprint not only the product but the process and the operators as well.

Figure 3.8 represents a record of a typical granulation batch done by an experienced operator on large Hobart mixer. The batch was stopped on the downslope of the derivative, that is, well inside the theoretical end-point region.

Figure 3.9 shows another batch made by the same operator. This time it is a power consumption trace, but again it extends beyond the peak of the derivative, and the end point thus can be deemed appropriate.

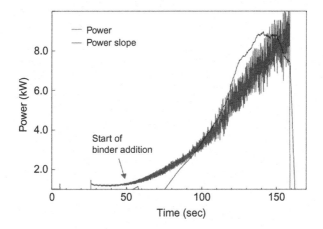

Figure 3.9 Another batch by the same operator (power consumption profile).

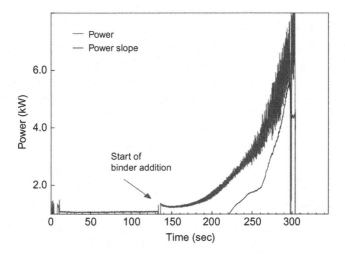

Figure 3.10 A batch by a novice operator (power consumption profile).

In the batch represented in Figure 3.10, a novice operator trainee has stopped the batch well before the peak of the derivative. This required a major adjustment of the tableting operation (force and speed) to produce tablets in the acceptable target range of properties (hardness and friability).

In another batch (see Fig. 3.11), the same novice operator has stopped the granulator, opened the lid, took a sample, and decided to granulate for another 10 sec. The peak of derivative was not reached at the stopping point.

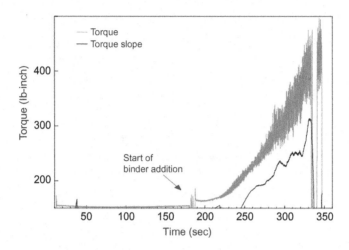

Figure 3.11 Another batch by an inexperienced operator (torque profile).

3.2.9 Emerging Technologies

A comprehensive review of wet granulation monitoring techniques can be found in Hansuld and Briens (2014). For some reason, the authors lump together measurements of power and torque, claiming that there is no significant difference between the techniques (an arguable statement not supported by evidence).

For the sake of completeness, we will mention some of the emerging technologies.

3.2.9.1 Acoustic Emission

Applicability of piezoelectric acoustic emission sensors to end-point determination has been studied since the beginning of this century (Whitaker, 2000; Hansuld et al., 2011). The technique is very promising, especially because it is noninvasive, sensitive, and relatively inexpensive. Briens et al. (2007) monitored vibration and sound waves emitted by 10 and 25-L granulators and observed a plateau at the granulation end point. They noted that the granulation phases were more clearly defined for the larger mixer.

Granulation process signatures obtained with acoustic transducer can be used to monitor changes in particle size, flow, and compression properties (Gamble et al., 2009). Daniher et al. (2008) used sound and vibration sensor in an attempt to detect the granulation end point.

They reported that acoustic methods, unlike vibration monitoring, allowed the optimum end point of granulation to be identified.

3.2.9.2 Near-Infrared Spectroscopy

Jørgensen (2004) showed that baseline corrected near-infrared (NIR) water absorbance and granulation impeller torque showed similarities during the water addition phase.

Use of a refractive NIR moisture sensor for end-point determination of wet granulation was described by several authors (Miwa et al., 2000; Otsuka et al., 2003; Muteki, 2012). Technological challenges are associated with this approach because the sensor can only measure the amount of water at the powder surface.

Near-infrared monitoring of the granulation process was attempted by researchers at many major pharmaceutical corporations with a modest success. In particular, the requirement of extensive calibration before use for each material and subsequent complex analysis requirement may dampen the initial enthusiasm with which the technique was greeted.

3.2.9.3 Focused Beam Reflectance Measurement

Focused beam reflectance measurement (FBRM) is a particle size determination technique based on a laser beam focusing in the vicinity of a sapphire window of a probe. The beam follows a circular path at speeds of up to 6 m/sec. When it intersects with the edge of a particle passing by a window surface, an optical collector records a backscatter signal. The time interval of the signal multiplied by the beam speed represents a chord length between two points on the edge of a particle. The chord length distribution (CLD) can be recalculated to represent either a number or volume-weighted PSD.

In many cases, when precision is more important than accuracy, CLD measurements are adequate to monitor dynamic changes in process parameters related to the particle size and shape, concentration, and rheology of fluid suspensions.

Several attempts were made to evaluate the use of FBRM particle size analyzer as a potential tool for granulation end-point determination (Ganguly and Gao, 2005). Dilworth et al. (2005) have compared power consumption, FBRM, and acoustic signals in a study of a wet granulation process in a Fielder PMA 200 mixer. It was found that these techniques were complementary, with the FBRM probe capable to follow median granule size growth even when the power consumption curve showed a plateau.

A major disadvantage of the FBRM method is that the measured CLD does not directly represent a PSD. Conversion of CLD to PSD is not straightforward and requires sophisticated mathematical software that is not easy to validate. Moreover, CLD depends on optical properties and shape of the particles, as well as the focal point position. The total number of counts measured is a function of both solids concentration and probe location.

3.2.9.4 Other Measurement Techniques

Microwave measurements using an open-ended coaxial probe (Gradinarsky et al., 2006) have been described as a useful measurement for in-line monitoring and control of wet granulation.

Electrical capacitance tomography (ECT) was also suggested as an in-line noninvasive technique (Rimpiläinen et al., 2011). However, this technique required extensive mixer modifications.

Raman spectroscopy was better suited to monitor API during wet granulation compared with NIR methods (Wikström, 2005).

Image probe was used by Watano (2001) to monitor growth of granule particle size and detect the end point of wet granulation. Watano et al. (2001) used a charge-coupled device (CCD) camera and high-energy xenon lighting system, with a fuzzy control system to control granule growth.

Talu et al. (2001) used stress fluctuation sensor to carry out a statistical analysis of a wet granulation process to determine the end point.

There are other ideas floating around, for example, use of neural network to describe and predict the behavior of the wet granulation (Watano, 1997).

Powder flow patterns in wet granulation can be studied using positron emission particle tracking (Saito, 2011).

A technique for measuring tensile strength of granules, in addition to power consumption measurement, to facilitate optimal end-point determination has been described by Betz et al. (2003).

Eventually, this and similar techniques can be used to validate various mathematical and statistical models of the process.

3.3 END-POINT DETERMINATION AND SCALE-UP: SEMI-EMPIRICAL METHODS

Several empirical scale-up methods have been suggested without much theoretical or practical justification. They are based on such process variables as blade tip speed and swept volume as measures of load or work of the main impeller.

3.3.1 Relative Swept Volume

Relative swept volume is defined as the volume swept out per second by the impeller blade divided by the volume of the mixer. The relative swept volume relates to the work of densification of the wet mass.

Relative swept volume has been suggested as a scale-up factor (Schaefer et al., 1986, 1987). This parameter is related to work of densification of wet mass done on the material and was studied extensively at various blade angles (Holm, 1987). Higher swept volume leads to higher temperature and denser granules. However, it was shown by Horsthuis et al. (1993) that the same relative swept volume did not result in the same end point (in terms of particle size distribution).

3.3.2 Impeller Tip Speed

Typical impeller tip speed of a high-shear granulator is about 8 to 12 m/sec. Higher speeds produce larger granules (Liu et al., 2009; Wade et al., 2015). Impeller tip speed corresponds to shear rate and has been used as a scale-up parameter in fluid mixing. Rekhi et al. (1996) worked with Fielder mixers and found that for a constant tip speed, successful scale-up is possible when liquid volume is proportional to the batch size and wet massing time ratio is inversely related to the ratio of impeller speeds. Other studies seem to support the constant tip speed suggestion (Aikawa et al., 2008).

However, for processing of lactose granulations in Gral mixers, it was shown by Horsthuis et al. (1993) that the same tip speed did not result in the same end point (in terms of particle size distribution). In this seminal and elegant work, Horsthuis and his colleagues from Organon in the Netherlands have studied granulation process in Gral mixers of 10-, 75-, and 300-L sizes. Comparing relative swept volume, blade tip speed, and Froude number, with respect to end-point determination (as expressed by the time after which there is no detectable change in particle size), they have concluded that only constant Froude numbers results in a comparable end point.

To clarify, a constant Froude number implies that the speed ratio between two processes is proportional to the square root of corresponding impeller diameter ratio (Eq. 3.5), and the constant tip speed requirement makes this relationship without the square root operation (Eq. 3.6):

$$\frac{N_2}{N_1} = \sqrt{\frac{D_2}{D_1}} \tag{3.5}$$

$$\frac{N_2}{N_1} = \frac{D_2}{D_1} \tag{3.6}$$

Comparison of constant tip speed, constant shear stress, and constant Froude number (Hassanpour, 2009) also showed that Froude numbers is a preferred variable to keep constant. And yet Sato et al. (2005) claimed that agitation power per unit vessel volume was better characterized by the tip speed rather than the Froude number, meaning that granule growth is mainly caused by the shear stress from the impeller blade.

However, at the same tip velocity, the wall of a smaller mixer will exert more energy on the granule than the wall of a larger mixer. This means that smaller machines must be run at a lower velocity compared to large machines in order to achieve a similar energy input.

Parikh (2010) presents an interesting comparison of impeller speed, impeller tip speed, and centrifugal acceleration at the two fixed impeller speeds of Fielder PMA high-shear mixer. Data on centrifugal acceleration reveal that there are higher compaction forces in smaller machines at the same level of tip speed.

According to Chitu et al. (2011), increasing impeller speed is found to generally reduce granule size, and the onset of breakage seems to occur for similar values of impeller tip speed. Granulating on a larger scale has shown that constant impeller tip speed offers good agreement in terms of mean granule size; however, granule size distribution seems to be scale dependent.

Although these findings only applied to specific materials, one of the main vendors of the high-shear granulators claimed that "experience indicates that the tip speed of the impellers should be held constant" (Rawland, M. 2007) in any scale-up operation and recommends the following procedure:

1. Calculate the tip speed S of the small scale impeller

$$S(m/s) = R_S \cdot 2\pi \cdot n/60 \qquad (3.2)$$

where R_S is the impeller blade radius and n is mixer speed.
Run the large scale mixer at speed of

$$60 \cdot S/(R_L \cdot 2\pi) \qquad (3.3)$$

where R_L is the impeller blade radius of the large-scale mixer.
2. The batch size should be kept approximately the same percent fill.
3. Calculate the binder at the same percent of dry mass.
4. Dry mix as needed (no rule; start with a similar time).
5. The volume of granulating fluid should be proportionally scaled up based on batch size.
6. Wet mass time is adjusted based on ratio of impeller speeds.
7. Use liquid addition nozzles that distribute binder in a "similar" manner.
8. Keep work constant to achieve a similar end point. Total impeller work is calculated as area under power consumption vs. time curve:

$$W = \sum P_t \Delta t \qquad (3.4)$$

Furthermore, divide W by total dry mass in a small-scale mixer to get a normalized impeller work and use it to calculate an appropriate target work value for the large mixer.

As you can see, this eclectic procedure is trying to combine equal impeller tip speed approach with the work equivalence based on power consumption. Unfortunately, there is no scientific justification for use of this semi-empirical method (except for, maybe, list items 7 and 8), which is why there is no peer-reviewed evidence of its success.

3.3.3 Experimental Design
The final goal of any granulation process is a solid dosage form, such as tablets. With so many variables involved in a granulation process, it is only

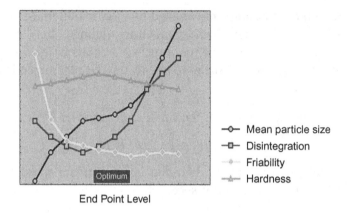

End Point Level

Figure 3.12 Wet granulation end point as a factor in tableting optimization.

natural to attempt to use the power of experimental design to arrive at an optimum response in terms of tablet properties (Iskandarani et al., 2001; Ogawa et al., 1994). Therefore, when optimizing a granulation process, the list of factors affecting tablet properties may include both the granulation end point and the tableting processing parameters, such as compression force or tablet press speed. In other words, granulation end point is but one factor in overall effort to make a better tableting product (Fig. 3.12).

In one of the most interesting works based on the experimental design approach, an attempt was made to find a statistical relationship between the major factors of both granulation and compaction unit operations, namely, granulation end point, press speed (dwell time), and compression force (Achanta et al., 1997). The resulting equation allowed optimization of such standard response parameters as tablet hardness, friability, and disintegration time. This study has also investigated the possibility of adjusting the tableting parameters to account for an inherent variability of a wet granulation process. Another good example of a comprehensive experimental design can be found in Woyna-Orlewicz and Jachowicz (2011).

Because the compressibility property of granulations is extremely sensitive to various processing parameters of wet granulation (Badawy et al., 2000), multivariate optimization of wet granulation may include hardness, disintegration, and ejection as response variables.

A more sophisticated approach called "regimen map" can offer several advantages over purely experimental design of experiments.

This quality by design approach is further facilitated by use of dimensionless parameters to reduce the number of variables required to quantify the process (Kayrak-Talay et al., 2013).

3.3.4 Examples

At low impeller speeds, high liquid addition rates or for some specific materials, the classic S shape of the power consumption curve may become distorted with a steep rise, leading into overgranulation. This case of a rapidly progressing process deserves special attention.

Zega et al. (1995) studied the process characterized by a rise of normalized power curve at a relatively constant rate in the region where the ratio of water to dry mass is 0.1 to 0.2 ("slope plateau"). A rapid increase in the slope of the power curve occurs in the region of the desired end point for the studied formulation. The desired end point (particle size, 110 to 160 microns) was reached for all mixer sizes (10-, 65-, and 250-L units) when the slope of power consumption exceeded the plateau level by a factor of 5 (empirical observation). Using this fact, the acceptable end point (target particle size of 135 microns) was first established on a 10-L Fielder and then scaled to a 65-L Fielder and a 250-L Diosna (Fig. 3.13).

Miyamoto et al. (1997) used orthogonal arrays experimental design to study a wet granulation process with explosive particle growth and to estimate the optimal response variables, such as yield, geometrical

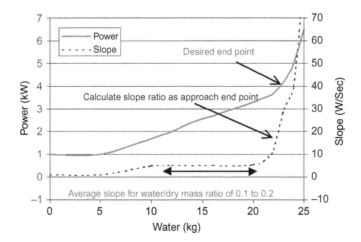

Figure 3.13 Use of relative level of derivative for rapidly granulating formulations.

mean of granules, uniformity of granule size, and granulation time. The technique was quite successful for two different sizes of Powrex mixer-granulators studied.

Campbell et al. (2011) studied high-shear wet granulation in a series of Fielder PMA mixers (25, 65, 150, and 300 L) and found that at optimal end points, the Froude numbers are linearly correlated with binder amount at inflection point of impeller power trace. This presented a good scale-up parameter if the condition of a constant tip speed was met.

CHAPTER 4

Recommended Procedures: Case Studies

4.1 CASE STUDY I: LEUENBERGER ET AL. (1979, 1983)

4.2 CASE STUDY II: LANDIN ET AL. (1996)

4.3 CASE STUDY III: FAURE ET AL. (1998)

4.4 CASE STUDY IV: LANDIN ET AL. (1999)

4.5 CASE STUDY V: FAURE ET AL. (1999)

4.6 CASE STUDY VI: HUTIN ET AL. (2004)

4.7 CASE STUDY VII: ANDRÉ ET AL. (2012)

4.1 CASE STUDY I: LEUENBERGER ET AL. (1979, 1983)

This case study is based on the pioneering studies conducted by Professor Hans Leuenberger at the University of Basel and Sandoz AG (Bier et al., 1979; Leuenberger et al., 1979; Leuenberger, 2001, Leuenberger and Betz, 2007).

Certain simplifying assumptions were used to construct the relevance list (Table 4.1). Most important were the assumptions that that there were no long-range particle interactions and no viscosity influence (and therefore, no Reynolds numbers).

Seven process parameters and one target variable (motor power consumption) represent the number $n = 8$ of the Π-theorem. There are $r = 3$ basic dimensions (M, L, and T). Per Buckingham's theorem, the process description can be reduced to the relationship between $m = n - r = 8 - 3 = 5$ mutually independent dimensionless groups.

Table 4.2 represents the dimensional matrix of this study. It was constructed as described above, with the rows listing the basic dimensions of mass (M), length (L), and time (T) and the columns specifying the physical quantities from the relevance list.

How to Scale-Up a Wet Granulation End Point Scientifically. DOI: http://dx.doi.org/10.1016/B978-0-12-803522-1.00004-0

Table 4.1 Relevance List for Case Study I

	Quantity	Symbol	Units	Dimensions
1	Power consumption	P	Watt	$M\,L^2\,T^{-3}$
2	Specific density	ρ	kg/m³	$M\,L^{-3}$
3	Blade diameter	d	m	L
4	Blade velocity	n	rev/s	T^{-1}
5	Binder amount	s	kg	M
6	Bowl volume	V_b	m³	L^3
7	Gravitational constant	g	m/s²	$L\,T^{-2}$
8	Bowl height	H	m	L

Table 4.2 Dimensional Matrix for Case Study I

	Core Matrix			Residual Matrix				
	ρ	d	n	P	s	V_b	g	H
Mass	1	0	0	1	1	0	0	0
Length	−3	1	0	2	0	3	1	1
Time	0	0	−1	−3	0	0	−2	0

Table 4.3 Unity and Residual Matrix for Case Study I

	Unity Matrix			Residual Matrix				
	ρ	d	n	P	s	V_b	g	H
M	1	0	0	1	1	0	0	0
3M + L	0	1	0	5	3	3	1	1
−T	0	0	1	3	0	0	2	0

L, *Length;* M, *mass;* T, *time.*

Now one needs to transform this dimensional matrix into a unity matrix as follows: to transform −3 in L-row column into zero, one linear transformation is required. The subsequent multiplication of the T-row by −1 transfers the −1 of the n-column to +1 (Table 4.3).

The dimensionless groups are formed from the columns of the residual matrix by dividing each element of the residual matrix by the column headers of the unity matrix, with the exponents indicated in the residual matrix.

The residual matrix contains five columns; therefore, five dimensionless Π groups (numbers) will be formed:

$$\Pi_0 = P/(r^1 \cdot d^5 \cdot n^3) = Np \quad \text{Newton (power) number}$$

$$\Pi_1 = q/(r^1 \cdot d^3 \cdot n^0) = q \cdot t/(V_p \cdot \rho) \quad \text{Specific amount of liquid,}$$

where $V_p \equiv$ Volume of particles, $q =$ binder addition rate, and $t =$ binder addition time.

$$\Pi_2 = t/(r^0 \cdot d^3 \cdot n^0) = (V_p/V_b)^{-1} \quad \text{Fractional particle volume}$$

$$\Pi_3 = g/(r^0 \cdot d^1 \cdot n^2) = Fr^{-1} \quad \text{Froude number}$$

$$\Pi_4 = h/(r^0 \cdot d^1 \cdot n^0) = h/d \quad \text{Ratio of lengths}$$

It is known from the dimensional analysis theory that all dimensionless Π groups are functionally related. The usual representation of this relationship is an expression of the form

$$\Pi_0 = f(\Pi_1, \Pi_2, \Pi_3, \Pi_4), \tag{4.1}$$

which states that the target quantity Π_0 is a function of all other Π groups.

Leuenberger and his coworkers made another simplifying assumption, namely, that the groups Π_2, Π_3, Π_4 are "essentially constant." In this case, the Π-space can be reduced to a relationship

$$\Pi_0 = f(\Pi_1), \tag{4.2}$$

that is, the value of Newton number Np at any point in the process is a function of the specific amount of granulating liquid.

The form of functional relationship f can be established experimentally.

Note that Π_1, the specific amount of liquid, is the binder liquid relative to batch size. Leuenberger and his group have established that the specific amount of binder liquid required to reach a desired endpoint (as expressed by the absolute value of Np and, by extension, in terms of net power consumption ΔP) is "scale-up invariable," that is, independent of the batch size (Fig. 4.1). This observation specified the functional dependence f and establishing rational basis for granulation scale-up.

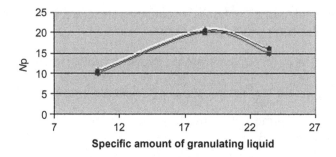

Figure 4.1 Newton power number as a function of the characteristic liquid quantity.

Experiments were performed with planetary mixers of five different batch sizes ranging from 3.75 to 60 kg. Nonviscous binder was mixed in with dry powder, and then liquid was added at a constant rate proportional to the batch size.

Power consumption was monitored, and the granulation process was stopped at different points after the peak of the signal derivative with respect to time. For each constant ratio of the granulation liquid quantity to a batch size, the end point was consistently similar irrespective of batch size and size of machine.

Note that if the binder addition rate is kept constant, then the amount of binder added will be proportional to time of binder addition, and the end point can be reproduced by maintaining the same time from peak of the derivative. It seems that before reaching the peak, the process may be influenced by different initial conditions. After the peak, one can start the stopwatch and stop the mixer after a predetermined amount of time has elapsed. An alternative to this method would be to monitor the level of power consumption that corresponds to the Newton power number.

Leuenberger's ideas relating to the use of power consumption for wet granulation end-point determination were tested and implemented by many researchers (Leuenberger et al., 1979; Holm et al., 1985a,b; Stamm and Paris, 1985; Werani, 1988; Terashita et al., 1990). These studies showed that for a constant rate of low-viscosity binder addition proportional to the batch size, the rate of change (slope or time derivative) of torque or power consumption curve is linearly related to the batch size for a wide spectrum of high shear and planetary mixers. In other words, the process end point, as determined by a characteristic amount of granulating liquid, is a theoretically and practically proven scale-up parameter for moving the product from laboratory to production mixers of different sizes and manufacturers.

Holm et al. (2001) have studied the effect of various processing factors on the correlation between power consumption and granule growth. They applied Leuenberger's methods to determine the end points in different experiments and found that such a correlation did indeed exist but was dependent on such factors as the impeller design, the impeller speed, and the type of binder. They compared the peak and level methods proposed by Leuenberger and concluded that it was possible to control the liquid addition by the level detection method (when the liquid addition is stopped at a predetermined level of power consumption). An alternative approach that involves an inflection point (peak of the signal derivative with respect to time) was not found to be applicable to all experiments.

4.2 CASE STUDY II: LANDIN ET AL. (1996)

This case study is discussed in Chapter 2, section 6, as a wet granulation example of dimensional analysis.

4.3 CASE STUDY III: FAURE ET AL. (1998)

A dimensional analysis approach was applied to planetary Hobart AE240 mixer with two interchangeable bowls (5 and 8.5 L). The following relevance list for the wet granulation process was suggested under the assumption of absence of chemical reaction and heat transfer (Table 4.4).

Table 4.4 Relevance List for Case Study III				
	Quantity	Symbol	Units	Dimensions
1	Net power	ΔP	Watt	$M\,L^2\,T^{-3}$
2	Wet mass bulk or specific density	ρ	kg/m^3	$M\,L^{-3}$
3	Impeller radius (or diameter)	d	M	L
4	Impeller speed	n	rev/s	T^{-1}
5	Granulation dynamic viscosity	η	Pa * s	$M\,L^{-1}\,T^{-1}$
6	Gravitational constant	g	m;/s^2	$L\,T^{-2}$
7	Height of granulation bed in the bowl	h	m	L

Note that, unlike Table 2.1 of the previous study, this list uses net power ΔP that was defined as motor power consumption under load minus the dry blending baseline level.

Several simplifying assumptions were made. One was that motor drive speed is proportional to the impeller blade speed. Another was that the ratio of characteristic lengths h/d was proportional to (and therefore can be replaced by) a fill ratio V_m/V_b, which was, in turn, assumed to be proportional to (and therefore could be replaced in the final equation by) the quantity $m/(\rho\ d^3)$. This is a preferred method of representing a fill ratio because the wet mass m is easier to measure than the height of the granulation bed in the bowl (see Section 6.1 for list of symbols). Furthermore, fill ratio was represented by $(\rho\ R_b{}^3/m)$ with the radius of the bowl R_b cubed used to replace the bowl volume V_b.

Application of dimensional analysis has reduced the number of variables from 7 dimensional to 4 dimensionless quantities that adequately describe the process: Np, ΨRe, Fr, and h/d.

Symbol ΨRe was introduced to denote a pseudo-Reynolds number as a measure of "wet mass consistency." It was calculated using torque values from mixer torque rheometer, which are proportional to kinematic viscosity $\nu = \eta/\rho$ rather than dynamic viscosity η required to compute true dimensionless Reynolds numbers. However, for a lack of impeller torque measurements that can be used to estimate dynamic viscosity, the authors used ΨRe in place of Re values.

A relationship of the form

$$N_p = 10^b(\Psi Re \cdot Fr \cdot \rho\ R_b^3/m)^a \qquad (4.3)$$

was postulated and converted into a log-log domain:

$$\text{Log}_{10}\ N_p = a \cdot \log_{10}(\Psi Re \cdot Fr \cdot \rho\ R_b^3/m) + b \qquad (4.4)$$

Constants a and b (slope and intercept) of this linear relation were found empirically with a good correlation (>0.92) between the observed and predicted numbers for both bowl sizes (Fig. 4.2).

Equation 4.4 can be interpreted to indicate that

$$\Delta P \sim \eta \cdot d^2 \cdot V_m/V_b, \qquad (4.5)$$

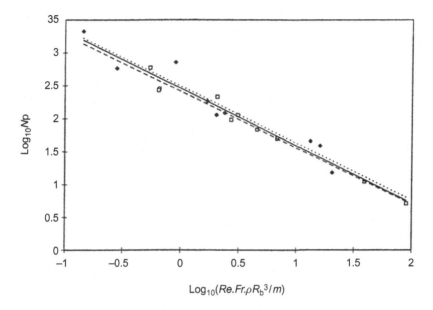

Figure 4.2 Regression graph for Case Study III.

This means that the net power consumption of the motor varies directly with the fill ratio, the wet mass viscosity, and the surface swept by the blades $(\sim d^2)$.

4.4 CASE STUDY IV: LANDIN ET AL. (1999)

Methodology presented in the previous case study was used to study granulation in planetary mixers Collette MP20, MP90, and MPH 200 under similar assumptions. The relevance list and dimensional matrix were the same as before, and torque measurements from torque rheometer were again used to calculate kinematic viscosity (instead of dynamic viscosity) in pseudo-Reynolds numbers.

Figure 4.3 represents the resulting the combined results from three mixers with bowl sizes 20, 90, and 200 liters. The regression line (Eq. 4.6) showed a pretty good fit to data ($r^2 > 0.95$).

$$\text{Log}_{10} \, N_p = a \cdot \log_{10}(\Psi Re \cdot Fr \cdot \rho \, R_b^3/m) + b \qquad (4.6)$$

Data from two other mixers in this study with bowl sizes 5 and 40 L produces lines that were significantly different from the first set of

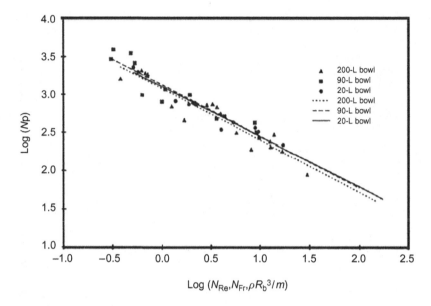

Figure 4.3 Regression graph for Case Study IV.

mixers. The authors explained this difference by "different flow patterns" and lack of geometrical similarity in the two groups of mixers.

4.5 CASE STUDY V: FAURE ET AL. (1999)

This dimensional analysis study was done on Collette Gral Mixers (8, 25, 75, and 600 L) and used similar relevance list and dimensionless numbers as the earlier studies. The challenge of applying this methodology to the scale-up in the Gral mixers was the lack of geometric similitude among the bowls at different scales. In addition, there was also lack of dynamic similarity because different mixer geometries resulted in different wall adhesion phenomenon that was partially relieved by using a polytetrafluoroethylene (PTFE) lining.

The resulting regression equation of the form

$$\text{Log}_{10} N_p = a \cdot \log_{10}(\Psi Re \cdot Fr \cdot \rho\, R_b^3/m) + b \tag{4.7}$$

was fitted to data from the 8-, 25-, and 75-L bowls with PTFE lining, and the 600-L bowl that did not require the lining (Fig. 4.4). The regression coefficient was $r^2 > 0.88$.

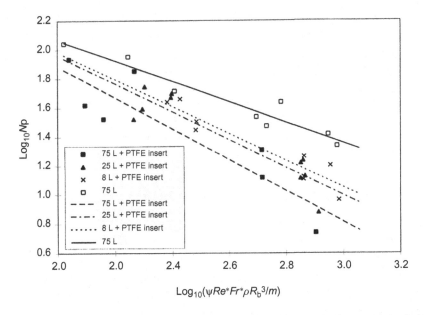

Figure 4.4 Regression graph for Case Study V. PTFE, Polytetrafluoroethylene. © 1999 Elsevier, reprinted with permission.

Table 4.5 Relevance List for Case Study VI

	Quantity	Symbol	Units	Dimensions
1	Power consumption	P	Watt	$M\,L^2\,T^{-3}$
2	Specific density	ρ	kg/m^3	$M\,L^{-3}$
3	Blade radius	d	m	L
4	Blade speed	n	rev/s	T^{-1}
5	Dynamic viscosity	η	Pa * s	$M\,L^{-1}\,T^{-1}$
6	Gravitational constant	g	m/s^2	$L\,T^{-2}$
7	Powder bed height	h	m	L
8	Blade length	l	m	L

4.6 CASE STUDY VI: HUTIN ET AL. (2004)

In this study, dimensional analysis was applied to a kneading process. Aoustin kneader with dual Z blades was instrumented for torque measurements and multiple runs of drug–cyclodextrin complexation were made at two scales (2.5 and 5 L).

In addition to other factors similar to those of the previous examples, the relevance list for this study (Table 4.5) includes the blade length as one of the crucial factors affecting the process.

When dimensional matrix is converted into unity matrix and dimensionless numbers are produced by the now familiar method, the blade length variable l becomes a denominator of a new dimensionless quantity d/l, so that the resulting regression equation is

$$N_p = b \left(\Psi Re \cdot Fr \cdot h/d \cdot d/l \right)^{-a} \qquad (4.8)$$

Experiments showed that the model fits data remarkably well ($r^2 > 0.99$).

4.7 CASE STUDY VII: ANDRÉ ET AL. (2012)

This study involved dimensional analysis of a mixing process on a planetary mixer. The first analysis considered mixing time, rotational and gyrational speeds, geometric parameters (vessel diameter, agitator height and diameter, bottom clearance, and solid height), material properties (densities and particle of the two mix components, solid fraction ratio), and speed of the agitator.

For fixed installation conditions, this 14-parametric dimensional space can be reduced to six principal variables (Table 4.6).

The dimensional matrix is given in Table 4.7.

Table 4.6 Relevance List for Case Study VII					
	Quantity	Symbol	Units		Dimensions
1	Mixing time	t	s		T
2	Density of first component	ρ_1	kg/m^3		M L^{-3}
3	Diameter of horizontal part of mixer	d	m		L
4	Gyrational speed of agitator	N_G	rev/s		T^{-1}
5	Rotational speed of agitator	N_R	rev/s		T^{-1}
7	Gravitational constant	g	m/s^2		L T^{-2}
8–14	Other variables (constant conditions)				

Table 4.7 Dimensional Matrix for Case Study VII						
	Core Matrix			Residual Matrix		
	N_G	ρ_1	d	t	N_R	g
Time	−1	0	0	1	−1	−2
Mass	0	1	0	0	0	0
Length	0	−3	1	0	0	1

Linear transformation creates a unity matrix and a residual matrix (Table 4.8):

Table 4.8 Unity and Residual Matrices for Case Study VII							
		Unity Matrix			Residual Matrix		
		N_G	ρ_1	d	t	N_R	g
$-T$	Time	1	0	0	-1	1	2
	Mass	0	1	0	0	0	0
$+3M$	Length	0	0	1	0	0	1
L, *Length;* M, *mass;* T, *time.*							

Application of Π-theorem reduces this six-parameter space to three-dimensional Π-groups (variables):

$$\Pi_0 = t \cdot N_G \tag{4.9}$$

$$\Pi_1 = N_R/N_G \tag{4.10}$$

$$\Pi_2 = (N_G)^2 \cdot d/g = \text{Froude number } Fr. \tag{4.11}$$

Thus, the target quantity t (mixing time) can be found from a relationship

$$F(t \cdot N_G, N_R/N_G, Fr) = 0 \tag{4.12}$$

The actual form of this relationship can be found empirically.

An earlier study by the same group of researchers (Delaplace et al., 2005 and 2007) performed dimensional analysis of planetary mixer for highly viscous fluids and arrived at a similar dimensionless function with Reynolds number instead of the Froude number.

CHAPTER 5

Practical Considerations for End-Point Scale-Up

5.1 HOW TO DEFINE AND DETERMINE THE END POINT

5.1.1 What is an End Point?

A wet granulation end point is usually defined by the formulator as a target particle size mean or distribution. It can also be defined empirically in rheological terms of wet mass density and viscosity, flowability, or tableting parameters (e.g., capping compression). Yet another way to define or quantify the end point is in terms of absorbed energy by calculating basic flow energy, specific energy, bulk density, and pressure drop or performing differential scanning calorimetry and effusivity measurements of the dry granules (Dave et al., 2012).

How to Scale-Up a Wet Granulation End Point Scientifically. DOI: http://dx.doi.org/10.1016/B978-0-12-803522-1.00005-2

There is no "correct" end point. There is only a desirable end point, a target defined by the formulator (Leuenberger et al., 2009).

The ultimate goal of monitoring a granulation process is to estimate the viscosity and density of the granules and perhaps to obtain an indication of the particle size mean and distribution. At any specific wet granulation end point, no matter how defined and measured, the set of rheological and statistical properties of the granules assume certain values. That means that the density and dynamic viscosity of the wet mass in any end-point condition are fixed at particular numerical values; the particle size distribution is fixed and can be calculated; and the only parameters that are left to vary are the process variables, namely, batch mass, impeller diameter and speed, and the geometry of the vessel used to arrive at this particular end point (Levin, 2013). Wet masses produced at the same end point (regardless of bowl and batch size, impeller speed, and moisture content) have been consistently shown to result in the same final dry granule size distribution, bulk density, flow, and mechanical strength.

It has been shown (Emori et al., 1997) that when you have reached the desired end point, the granule properties and the subsequent tablet properties are very similar regardless of the granulation processing factors, such as impeller or chopper speed or binder addition rate. We postulate that it does not matter what process conditions were used to achieve any particular end point. There seems to exist a "principle of equifinality" that states: "An end point is an end point regardless of how it was obtained."

Conversely, if you follow a different processing pathway, you may never arrive to the same end point unless you use some sort of precision monitoring of crucial process parameters and stop the process when these parameters reach predetermined values because each different end point has its own set of granulation properties. These process parameters are best characterized by certain numerical values of the dimensionless variables describing the process. It is true that you do not need dimensional analysis to define and reproduce an end point. For this task, you could successfully use such semi-empirical methods as maintaining the same blade tip speed or area under swept volume–time curve. You need dimensional analysis only when your eventual goal is to reproduce the end point with a different set of processing factors or on a different mixer. When a particular end point is reached, it can be characterized by a set of values of dimensionless variables, and, because the variables are dimensionless, they are independent of scale.

5.1.2 How to Find an Optimal End Point

It is advisable to run a trial batch on a small-scale granulator at a fixed speed with a predetermined method of binder addition (e.g., add water continuously at a fixed rate to a dry mix with a water-soluble binding agent).

Before adding the liquid, measure the baseline level of motor power consumption P_o or impeller torque τ_o at the dry mix stage.

During the batch, stop the process frequently to take samples and, for each sample, note the end-point values of power consumption P_e or impeller torque τ_e. For each of these "end points," measure the resulting wet mass density and viscosity (if using a viscous binder) and determine a particle size distribution. As a result, you will be able to obtain some data that will relate the spectrum of "end-point parameters" listed earlier to the processing variables in terms of net motor power consumption

$$\Delta P_m = (P_e - P_o) \tag{5.1}$$

or net impeller power consumption

$$\Delta P_i = 2\pi * n * (\tau_e - \tau_o), \tag{5.2}$$

where n is the impeller speed (dimension T^{-1}).

5.2 HOW TO REPRODUCE THE END POINT

5.2.1 Same Mixer and Processing Parameters

After the desired end point is defined, it can be reproduced by stopping the batch at the same level of net power consumption ΔP (for the same mixer, formulation, speed, batch size, and amount of granulating liquid). If impeller torque measurements are available, it is best to establish the net power consumption ΔP_i (Eq. 5.2) of the impeller rather than the net power consumption ΔP_m (Eq. 5.1) of the motor.

In any case, the target value is determined as the end-point power consumption P minus the baseline power consumption P_0 (empty mixer or dry mix value). The granulator should be stopped when reach the previously acceptable value of ΔP is reached.

To reiterate, to reproduce an end point, it is sometimes sufficient to monitor power of the impeller (or the motor) and stop when a predefined net level of the signal is reached. If, however, any of the processing variables or the rheological definition of the end point has changed, a more sophisticated approach is required, as described below.

5.2.2 Same Mixer; Different Processing Parameters

Wet granulation process can be quantified by dimensionless Newton power number Np that will assume a certain numerical value for every state (condition) of the granulating process. Under fixed processing conditions, Np will be proportional to net power consumption ΔP for any end point (defined, in part, by wet mass density).

At that point and for any future scale-up or technology transfer work, it is advisable to calculate the Newton power number from the net power consumption ΔP, density, speed, and blade diameter as follows:

$$Np = \Delta P/(\rho \cdot n^3 \cdot d^5) \tag{5.3}$$

This value will be used in this and the following sections.

Changes in formulation (but not in viscosity of binder), processing speed, and amount or rate of granulating liquid necessitate recalculation of ΔP to adjust for the changes. The end-point ΔP will differ from a previously established value. What should not change is the Newton power number Np for the desired end point, and after you know the end-point value of Np, you can use it to backcalculate the stopping value of the end-point ΔP from Equation 5.4:

$$\Delta P = Np \cdot \rho \cdot n^3 \cdot d^5 \tag{5.4}$$

In other words, if you have established an end- in terms of some net impeller or motor power ΔP and would like to reproduce this end point on the same mixer at a different speed or wet mass density, calculate Newton power number Np (Eq. 5.3) from the given net impeller power ΔP, impeller speed n, blade radius d, and wet mass density

ρ (assuming the same batch size) and then recalculate the target ΔP with the changed values of speed n or wet mass density ρ. When you know the new target ΔP, stop the granulator when your monitor says that the target value is reached.

5.2.3 Different But Similar Mixer

We already know that for every blend and a fixed set of values for processing factors (e.g., mixer geometry, blade speed, powder volume, amount and method of addition of granulating liquid), a wet granulation process state (end point) is completely characterized by rheological properties of the wet mass (density, viscosity), which are, in turn, a function of particle size, shape, and other properties. The process can be quantified with the help of dimensionless Newton power number Np and other dimensionless numbers that will assume certain numerical values for every state (condition) of the granulate.

Reproducing the wet granulation end point in a different mixer (under the assumption of similitude) is already a scale-up task even if the batch size did not change significantly. This task requires full application of dimensional analysis as will be described later.

5.3 HOW TO SCALE-UP THE END POINT

We recommend a scientifically proven method of dimensional analysis in which Froude, Reynolds, and Newton power numbers play a pivotal role as discussed in previous chapters. Following the examples discussed in the case studies, you can combine the results obtained at different end points of the test batch or from different batches or mixer scales (assuming geometric similarity).

After the optimal end point is chosen (in terms of granulate properties, related processing variables, and corresponding dimensional numbers), run dimensional analysis to generate dimensional groups, as explained in previous chapters.

The objective is to establish net impeller power ΔP from the target Newton power number Np as described earlier. To get the required Np, you can use the dimensionless regression equation of the form:

$$Np = b \cdot (Re \cdot Fr \cdot h/d)^a, \tag{5.5}$$

where h is bowl height, h/d is fill ratio, "a" and "b" are the slope and intercept of the (empirical, mixer, and powder-specific) regression equation based on end-point parameters.

Input required for dimensional analysis of a conventional high-shear mixer-granulator run is listed as follows:

- Blade radius (impeller diameter), m
- Powder weight, kg
- Liquid density, kg/m^3
- Rate of liquid addition, mL/min

See examples below showing how to calculate fill ratio, wet mass viscosity, Newton power number, Froude number, and Reynolds number to generate values for Equation 5.5.

Note that different mixer types and vessel and blade geometry contribute to the differences in absolute values of the signals. However, the signal profile of a given granulate composition in a high-shear mixer is very similar to the one obtained in a planetary mixer.

For accuracy, in power number Np calculations, the power of the load on the impeller rather than the mixer motor should be used. Unlike power consumption of the impeller (based on torque measurements), the baseline for motor power consumption does not stay constant and can change significantly with load on the impeller, mixer condition, or motor efficiency. This may present inherent difficulties in using power meters instead of torque. Torque, of course, is directly proportional to power drawn by the impeller (the power number can be determined from the torque and speed measurements) and has a relatively constant baseline.

5.3.1 How to Calculate Fill Ratio

You can calculate fill ratio, given powder weight, granulating liquid density ($1000 \, kg/m^3$ for water), rate of liquid addition, time interval for liquid addition, and bowl volume V_b.

The calculations are performed using the idea that the fill ratio h/d (wet mass height to blade diameter) is proportional to V/V_b, and wet mass volume V can be computed as

$$V = m/\rho, \qquad (5.6)$$

where m is the mass (weight) of the wet mass and ρ is the wet mass density.

Now, the weight of the wet mass is computed as the weight of powder plus the weight of added granulating liquid. The latter, of course, is calculated from the rate and duration of liquid addition and the liquid density.

For example, using powder weight $m = 1.2\,\text{kg}$, liquid density $\eta L = 1000\,\text{kg/m}^3$, bowl volume $V_b = 4\,\text{l}$, rate of liquid addition $= 21\,\text{mL/min}$, and time of liquid addition $= 270\,\text{s}$, the fill ratio is calculated to be 0.539.

If liquid is added at once, then, instead of rate of liquid addition and time interval for liquid addition, use liquid volume of 94.5 mL to get the same result.

5.3.2 How to Calculate Froude Number
Given the speed and blade diameter, the calculation of the Froude number is straightforward:

$$Fr = n^2 d/g \qquad (5.7)$$

5.3.3 How to Calculate Wet Mass Viscosity
Wet mass viscosity is required for calculation of Reynolds number in Equation. 4.3. Given net impeller power ΔP, blade radius R, and impeller speed n, wet mass viscosity η can be calculated using the following equations:

$$\Delta P = 2\pi\,\Delta\tau \cdot n \qquad (5.8)$$

$$\eta = \varphi \cdot \Delta\tau/(n \cdot d^3) \qquad (5.9)$$

where $\Delta\tau$ is the net torque required to move wet mass, n is the speed of the impeller, d is the blade diameter, and φ is mixer specific "viscosity factor" relating torque and dynamic viscosity. (Note: φ can be established empirically by running a mixer with water that has dynamic viscosity $\eta = 1$ and measuring torque.) Alternatively, you can use impeller torque τ as a measure of kinematic viscosity and use it to obtain a nondimensionless "pseudo-Reynolds" number based on the so-called "mix consistency" measure, that is, the end-point torque, as described in the case studies.

For example, if the "viscosity factor" φ for your specific mixer was found to be 0.000625, then for $n = 400$ RPM, $R = 0.4$ m, and net impeller power $= 385$ W, wet mass viscosity is calculated to be 0.06597 Pa · s.

5.3.4 How to Calculate Reynolds Number

Given wet mass density ρ, wet mass viscosity η, setup speed n, and blade radius or diameter d, you can calculate the Reynolds number Re as

$$Re = d^2 n \rho / \eta \tag{5.10}$$

or the "pseudo-Reynolds" number if you use a torque rheometer,

$$\Psi Re = d^2 n \, \rho / \tau \tag{5.11}$$

5.3.5 How to Find Slope "a" and Intercept "b"

After you have established fill ratio, and the Newton power, Froude, and Reynolds numbers for a range of end points, the data required for further analysis are obtained by running a simple linear regression. The objective is to estimate the coefficients "a" and "b" of Equation 5.5.

These coefficients become slope "a" and intercept "b" of the linearized equation

$$\log N_p = \log b + a \cdot \log (Re \cdot Fr \cdot h/d) \tag{5.12}$$

And, inversely, after the regression line is established, you can calculate Newton power number Np (which is the target quantity for scale up) and net power ΔP (which can be observed in real time as a true indicator of the target end point) for any point on the line (scale-up path).

5.3.6 How to Calculate Newton Power Number

Normally, you would want to calculate Newton power number Np from net power consumption (Eq. 5.3), given Net impeller power consumption ΔP, impeller speed n, blade radius R, and wet mass density ρ. This may be useful if you have established an end point in terms of some net impeller power ΔP and would like to reproduce this end point

Practical Considerations for End-Point Scale-Up   63

on the same mixer at a different speed or wet mass density (keeping the target Newton power number Np).

For example, using the preceding example a value of 350 W for net impeller power, the end-point values are calculated to be:

Reynolds number = 28,809.5
Froude number = 0.779
Target net impeller power = 350.0 (W)
Newton power number = 62.6

5.3.7 How to Calculate ΔP from the Dimensionless Equation

The wet granulation process can be quantified by dimensionless Newton power number Np that will assume a certain numerical value for every state (condition) of the granulating process. Under fixed processing conditions, Np will be proportional to net power consumption ΔP for any end point (defined, in part, by wet mass density).

The target quantity for reproducible end point and for scale-up is net impeller power ΔP. It should be calculated and monitored in real time, and the granulator should be stopped when the target number is reached.

The net impeller power number can be calculated, given wet mass density ρ, wet mass viscosity η, fill ratio h/d (roughly proportional to $m \cdot V_b/\rho$), setup speed n, and blade radius R, as well as the slope "a" and intercept "b" from the regression in Equation 5.12.

For example, for $\rho = 600$ kg/m³, $\eta = 0.0021$ Pa $\cdot$ s (at viscosity factor $\varphi = 7.40E\text{-}04$), $h/d = 0.539$, $n = 500$ RPM, $R = 0.11$ m, $a = -0.72$, $b = 4.5$, the end-point values are calculated to be:

Reynolds number = 28,809.5
Froude number = 0.779
Newton power number = 36.3
Target net impeller power = 203.2

To reach the same process state at any scale, stop the mixer when Np or a corresponding ΔP has reached a predetermined end-point value.

CHAPTER 6

List of Symbols and Dimensions

a, b	Slope and intercept of a regression equation
d	Impeller (blade) diameter or radius (m); dimensional units [L]
g	Gravitational constant (m/s^2); dimensional units [LT^{-2}]
h	Height of granulation bed in the bowl (m); dimensional units [L]
H	Bowl height (m); dimensional units [L]
l	Blade length (m); dimensional units [L]
n	Impeller speed (revolutions/s); dimensional units [T^{-1}]
P	Power required by the impeller or motor (W = J/s); dimensional units [ML^{2}T^{-5}]
R_b	Radius of the bowl (m); dimensional units [L]
q	Binder liquid addition rate
s	Amount of granulating liquid added per unit time (kg); dimensional units [M]
t	Binder addition time (s); dimensional units [T]
V_p	Particle volume (m^3); dimensional units [L^3]
V_m	Wet mass volume (m^3); dimensional units [L^3]
V_b	Bowl volume (m^3); dimensional units [L^3]
w	Wet mass; dimensional units [M]
ρ	Specific density of particles (kg/m^3); dimensional units [M L^{-3}]
$\nu = \eta/\rho$	Kinematic viscosity (m^2/s); dimensional units [L^{2}T^{-1}]
η	Dynamic viscosity (Pa · s); dimensional units [M L^{-1} T^{-1}]
τ	Torque (N-m); dimensional units [M L^2 T^{-2}]. Note: Torque has the same dimensions as work or energy.
$\varphi = \eta \cdot n \cdot d^3/\Delta\tau$	Dimensionless "viscosity factor" relating net torque $\Delta\tau$ and dynamic viscosity η
$Fr = n^2\, d/g$	Froude number. It relates the inertial stress to the gravitational force per unit area acting on the material. It is a ratio of the centrifugal force to the gravitational force.
$N_p = P/(\rho\, n^3\, d^5)$	Newton (power) number. It relates the drag force acting on a unit area of the impeller and the inertial stress.
$Re = d^2\, n\, \rho/\eta$	Reynolds number. It relates the inertial force to the viscous force.
$\Psi Re = d^2\, n\, \rho/\tau$	"Pseudo-Reynolds number" (m^3/s); dimensional units [L^{-3} T]. If end-point torque values are obtained from mixer torque rheometer and are used to calculate the Reynolds number, the result is a dimensional number described as "wet mass consistency" number. Note: This variable physically is a reciprocal of volume flow rate.

How to Scale-Up a Wet Granulation End Point Scientifically. DOI: http://dx.doi.org/10.1016/B978-0-12-803522-1.00006-4

REFERENCES

Achanta, A.S., Adusumilli, P.S., James, K.W., 1997. Endpoint determination and its relevance to physicochemical characteristics of solid dosage forms. Drug Dev. Ind. Pharm. 23 (6), 539–546.

Aikawa, S., Fujita, N., Myojo, H., et al., 2008. Scale-up studies on high shear wet granulation process from mini-scale to commercial scale. Chem. Pharm. Bull. (Tokyo) 56 (10), 1431–1435.

André, C., Demeyre, J.F., Gatumel, C., et al., 2012. Dimensional analysis of a planetary mixer for homogenizing of free flowing powders: mixing time and power consumption. Chem. Eng. J. 198–199, 371–378.

Badawy, S.I.F., Menning, M.M., Gorko, M.A., Gilbert, D.L., 2000. Effect of process parameters on compressibility of granulation manufactured in a high-shear mixer. Int. J. Pharm. 198 (1), 51–61.

Belohlav, Z., Brenková, L., Hanika, J., et al., 2007. Effect of drug active substance particles on wet granulation process. Chem. Eng. Res. Des. 85 (7), 974–980.

Betz, G., Burgin, P., Leuenberger, H., 2003. Power consumption profile analysis and tensile strength measurements during moist agglomeration. Int. J. Pharm. 252 (1–2), 11–25.

Betz, G., Bürgin, P.J., Leuenberger, H., 2004. Power consumption measurement and temperature recording during granulation. Int. J. Pharm. 272 (1–2), 137–149.

Bier, H., Leuenberger, H., Sucker, H., 1979. Determination of the uncritical quantity of granulating liquid by power measurements on planetary mixers. Pharm. Ind. 41 (4), 375–380.

Briens, L., Daniher, D., Tallevi, A., 2007. Monitoring high-shear granulation using sound and vibration measurements. Int. J. Pharm. 331 (1), 54–60.

Buckingham, E., 1914. On physically similar systems; illustrations of the use of dimensional equations. Phys. Rev. 4, 345–376.

Campbell, G.A., Clancy, D.J., Zhang, J.X., et al., 2011. Closing the gap in series scale up of high shear wet granulation process using impeller power and blade design. Powder Technol. 205 (1–3), 184–192.

Cavinato, M., Bresciani, M., Machin, M., et al., 2010a. Formulation design for optimal high-shear wet granulation using on-line torque measurements. Int. J. Pharm. 387 (1–2), 48–55.

Cavinato, M., Bresciani, M., Machin, M., et al., 2010b. The development of a novel formulation map for the optimization of high shear wet granulation. Chem. Eng. J. 164 (2–3), 350–358.

Chitu, T.M., Oulahna, D., Hemati, M., 2011. Wet granulation in laboratory scale high shear mixers: effect of binder properties. Powder Technol. 206 (1–2), 25–33.

Cliff, M.J., 1990. Granulation end point and automated process control of mixer-granulators: part I. Pharm. Tech. 4, 112–132.

Corvari, V., Fry, W., Seibert, W., Augsburger, L., 1992. Instrumentation of a high-shear mixer: evaluation and comparison of a new capacitive sensor, a watt meter, and a strain-gage torque sensor for wet granulation. Pharm. Res. 9, 1525–1533.

Daniher, D., Briens, L., Tallevi, A., 2008. End-point detection in high-shear granulation using sound and vibration signal analysis. Powder Technol. 181 (2), 130–136.

Dave, R., Wu, S., Contractor, L., 2012. To determine the end point of wet granulation by measuring powder energies and thermal properties. Drug Dev. Ind. Pharm. 38 (4), 439–446.

Delaplace, G., Guerin, R., Leuliet, J.C., 2005. Dimensional analysis for planetary mixer: Modified power and Reynolds numbers. AIChE. J. 51 (12), 3094−3100.

Delaplace, G., Thakur, R.K., Bouvier, L., et al., 2007. Dimensional analysis for planetary mixer: Mixing time and Reynolds numbers. Chem. Eng. Sci. 62 (5), 1442−1447.

Dilworth, S., Mackin, L., Weir, S., et al., 2005. In-line techniques for end-point determination in large scale high shear wet granulation. J. Pharm. Pharmacol. 57, S25.

Elliott, T., 1993. Efficiency, reliability of drive systems continue to improve. Power 137 (2), 33−41.

Emori, H., Sakuraba, Y., Takahashi, K., et al., 1997. Prospective validation of high-shear wet granulation process by wet granule sieving method. II. Utility of wet granule sieving method. Drug Dev. Ind. Pharm. 23 (2), 203−215.

Faure, A., Grimsey, I.M., Rowe, R.C., et al., 1998. A methodology for the optimization of wet granulation in a model planetary mixer. Pharm. Dev. Technol. 3, 413−422.

Faure, A., Grimsey, I.M., Rowe, R.C., et al., 1999a. Applicability of a scale-up methodology for wet granulation processes in Collette Gral high shear mixer-granulators. Eur. J. Pharm. Sci. 8 (2), 85−93.

Faure, A., Grimsey, I.M., Rowe, R.C., et al., 1999b. Process control in a high shear mixer-granulator using wet mass consistency: the effect of formulation variables. J. Pharm. Sci. 88 (2), 191−195.

Faure, A., York, P., Rowe, R.C.C., 2001. Process control and scale-up of pharmaceutical wet granulation processes: a review. Eur. J. Pharm. Biopharm. 52 (3), 269−277.

Food and Drug Administration, 1995. SUPAC-IR: Immediate Release Solid Oral Dosage Forms.

Food and Drug Administration, 1997. SUPAC-MR: Modified Release Solid Oral Dosage Forms.

Food and Drug Administration, 2004. Guidance for Industry Guidance for Industry PAT—A Framework for Innovative Pharmaceutical. (September), 1−19.

Fink, D., Beaty, H., 1993. Standard Handbook for Electrical Engineers, 13th ed. McGraw-Hill, New York.

Fry, W., Stagner, W., Wichman, K., 1987. Computer-interfaced capacitive sensor for monitoring the granulation process 2: system response to process variables. J. Pharm. Sci. (10), 30−41.

Gamble, J.F., Dennis, A.B., Tobyn, M., 2009. Monitoring and end-point prediction of a small scale wet granulation process using acoustic emission. Pharm. Dev. Tech. 14 (3), 299−304.

Ganguly, S., Gao, J.Z., 2005. Application of on-line Focused Beam Reflectance Measurement Technology in high shear wet granulation. AAPS J. 7.

Gradinarsky, L., Brage, H., Lagerholm, B., et al., 2006. In situ monitoring and control of moisture content in pharmaceutical powder processes using an open-ended coaxial probe. Meas. Sci. Technol. 17 (7), 1847−1853.

Hancock, B.C., York, P., Rowe, R.C., Parker, M.D., 1991. Characterization of wet masses using a mixer torque rheometer: 1. Effect of instrument geometry. Int. J. Pharm. 76 (3), 239−245.

Hansuld, E.M., Briens, L., McCann, J.A., Sayani, A., 2009. Audible acoustics in high-shear wet granulation: application of frequency filtering. Int. J. Pharm. 378 (1), 37−44.

Hansuld, E.M., Briens, L., Sayani, A., McCann, J.A., 2012. Monitoring quality attributes for high-shear wet granulation with audible acoustic emissions. Powder Technol. 215, 117−123.

Hansuld, E.M., Briens, L., 2014. A review of monitoring methods for pharmaceutical wet granulation. Int. J. Pharm. 472 (1), 192−201.

Hariharan, M., Mehdizadeh, E., 2002. The use of mixer torque rheometry to study the effect of formulation variables on the properties of wet granulations. <http://informahealthcare.com/doi/abs/10.1081/DDC-120002841> (accessed 19.05.15).

Hassanpour, A., Kwan, C.C., Ng, B.H., et al., 2009. Effect of granulation scale-up on the strength of granules. Powder Technol. 189 (2), 304–312.

Hirzel, J., 1992. Understanding premium-efficiency motor economics. Plant Eng. 7, 75–78.

Holm, P., 1987. Effect of impeller and chopper design on granulation in a high speed mixer. Drug Dev. Ind. Pharm. 13, 1675–1681.

Holm, P., 1997. High shear mixer granulators. In: Parikh, D.M. (Ed.), Handbook of Pharmaceutical Granulation Technology. Marcel Dekker, New York.

Holm, P., Jungersen, O., Schaefer, T., Kristensen, H., 1983. Granulation in high speed mixers. Part I: effect of process variables during kneading. Pharm. Ind. 45, 806–811.

Holm, P., Schaefer, T., Kristensen, H.G., 1985a. Granulation in high-speed mixers. Part V. Power consumption and temperature changes during granulation. Powder Technol. 43 (3), 213–223.

Holm, P., Schaefer, T., Kristensen, H.G., 1985b. Granulation in high-speed mixers. Part VI. Effects of process conditions on power consumption and granule growth. Powder Technol. 43 (3), 225–233.

Holm, P., Schaefer, T., Larsen, C., 2001. End-point detection in a wet granulation process. Pharm. Dev. Technol. 6 (2), 181–192.

Horsthuis, G.J.B., Van Laarhoven, J.A.H., Van Rooij, R.C.B.M., et al., 1993. Studies on upscaling parameters of the Gral high shear granulation process. Int. J. Pharm. 92 (1–3), 143–150.

Hutin, S., Chamayou, A., Avan, J.L., et al., 2004. Analysis of a kneading process to evaluate drug substance–cyclodextrin complexation. Pharm. Tech. 28, 112–123.

Iskandarani, B., Shiromani, P.K., Clair, J.H., 2001. Scale-up feasibility in high-shear mixers: determination through statistical procedures. Drug Dev. Ind. Pharm. 27 (7), 651–657.

Jørgensen, P., Pedersen, J.G., Jensen, E.P., Esbensen, K.H., 2004. On-line batch fermentation process monitoring (NIR)—introducing 'biological process time'. J. Chemometrics 18 (2), 81–91.

Kayrak-Talay, D., Dale, S., Wassgren, C., Litster, J., 2013. Quality by design for wet granulation in pharmaceutical processing: assessing models for a priori design and scaling. Powder Technol. 240, 7–18.

Knight, P.C., Instone, T., Pearson, J.M.K., Hounslow, M.J., 1998. An investigation into the kinetics of liquid distribution and growth in high shear mixer agglomeration. Powder Technol. 97 (3), 246–257.

Kristensen, H.G., 1996. Particle agglomeration in high shear mixers. Powder Technol. 88 (3), 197–202.

Kumar, A., Gernaey, K.V., Beer, T., De, Nopens, I., 2013. Model-based analysis of high shear wet granulation from batch to continuous processes in pharmaceutical production—a critical review. Eur. J. Pharm. Biopharm. 85 (3 part B), 814–832.

Laicher, A., Profitlich, T., Schwitzer, K., Ahlert, D., 1997. A modified signal analysis system for end-point control during granulation. Eur. J. Pharm. Sci. 5 (1), 7–14.

Landin, M., 1996. The effect of batch size on scale-up of a pharmaceutical granulation in a fixed bowl mixer granulator. Int. J. Pharm. 134 (1–2), 243–246.

Landín, M., Rowe, R.C., York, P., 1995. Characterization of wet powder masses with a mixer torque rheometer. 3. Nonlinear effects of shaft speed and sample weight. J. Pharm. Sci. 84 (5), 557–560.

Landin, M., York, P., Cliff, M.J., et al., 1996. Scale-up of a pharmaceutical granulation in fixed bowl mixer-granulators. Int. J. Pharm. 133 (1–2), 127–131.

Landín, M., York, P., Cliff, M.J., Rowe, R.C., 1999. Scaleup of a pharmaceutical granulation in planetary mixers. Pharm. Dev. Technol. 4 (2), 145–150.

Leuenberger, H., 1983. Scale-up of granulation processes with reference to process monitoring. Acta Pharm. Technol. 29 (4), 274–280.

Leuenberger, H., 2001. New trends in the production of pharmaceutical granules: batch versus continuous processing. Eur. J. Pharm. Biopharm. 52 (3), 289–296.

Leuenberger, H., Betz, G., 2007. Granulation process control—production of pharmaceutical granules: the classical batch concept and the problem of scale-up. Granulation. Handbook of Powder Technology. Elsevier, St. Louis, pp. 705–733.

Leuenberger, H., Bier, H., Sucker, H., 1979. Theory of the granulating-liquid requirement in the conventional granulation process. Pharm. Tech. 6, 61–68.

Leuenberger, H., Puchkov, M., Krausbauer, E., Betz, G., 2009. Manufacturing pharmaceutical granules: is the granulation end-point a myth? Powder Technol. 189 (2), 141–148.

Levi, M., Zlokarnik, M., 2011. Dimensional analysis of the tableting process. In: Levin, M. (Ed.), Pharmaceutical Process Scale-Up, third ed. Informa Healthcare Ltd. UK, London.

Levin, M., 2013. Wet granulation: end-point determination and scale-up, in: Encyclopedia of Pharmaceutical Science and Technology, fourth ed., CRC Press, New York, pp. 3854–3871.

Liu, L., Levin, M., Sheskey, P., 2009. Process development and scale-up of wet granulation by the high shear process. In: Developing Solid Oral Dosage Forms: Pharmaceutical Theory and Practice, Elsevier, Amsterdam, pp. 667–700.

Mackaplow, M.B., Rosen, L.A., Michaels, J.N., 2000. Effect of primary particle size on granule growth and endpoint determination in high-shear wet granulation. Powder Technol. 108 (1), 32–45.

Merrifield, C.W., 1870. The experiments recently proposed on the resistance of ships. Trans. Inst. Nav. Arch. 11, 80–93.

Miwa, A., Yajima, T., Itai, S., 2000. Prediction of suitable amount of water addition for wet granulation. Int. J. Pharm 195 (1–2), 81–92.

Miyamoto, Y., Ogawa, S., Miyajima, M., et al., 1997. An application of the computer optimization technique to wet granulation process involving explosive growth of particles. Int. J. Pharm. 149 (1), 25–36.

Muteki, K., Yamamoto, K., Reid, G.L., Krishnan, M., 2011. De-risking scale-up of a high shear wet granulation process using latent variable modeling and near-infrared spectroscopy. J. Pharm. Innov. 6 (3), 142–156.

Nichols, G., Byard, S., Bloxham, M.J., et al., 2002. A review of the terms agglomerate and aggregate with a recommendation for nomenclature used in powder and particle characterization. J. Pharm. Sci. 91 (10), 2103–2109.

Ogawa, S., Kamijima, T., Miyamoto, Y., et al., 1994. A new attempt to solve the scale-up problem for granulation using response surface methodology. J. Pharm. Sci. 83 (3), 439–443.

Otsuka, M., Mouri, Y., Matsuda, Y., 2003. Chemometric evaluation of pharmaceutical properties of antipyrine granules by near-infrared spectroscopy. AAPS Pharm. Sci. Tech. 4 (3), E47.

Parikh, D.M. (Ed.), 2010. Handbook of Pharmaceutical Granulation Technology. third ed. Taylor & Francis, Boca Raton, FL.

Parker, M., Rowe, R., Upjohn, N., 1990. Mixer torque rheometry: a method for quantifying the consistency of wet granulations. Pharm. Tech. Int. 2, 50–64.

Rawland, M., Scale-up for PMA High Shear Granulators. White Paper. <http://www.scribd.com/doc/33402031/PMA-Scale-up-White-Paper#scribd> (accessed 12.12.14.).

Rayleigh, L., 1915. The principle of similitude. Nature 95 (66), 591.

Record, P., 1979. Practical experience with high-speed pharmaceutical mixer/granulators. Manuf. Chem. Aerosol News. 11, 65–67.

Rekhi, G.S., Caricofe, R.B., Parikh, D.M., Augsburger, L.L., 1996. A new approach to scale-up of a high-shear granulation process. Pharm. Technol. October.

Reynolds, O., 1883. An experimental investigation of the circumstances which determine whether the motion of water shall be direct or sinuous, and of the law of resistance in parallel channels. Philos. Trans. R. Soc. London 174, 935–982.

Rimpiläinen, V., Poutiainen, S., Heikkinen, L.M., et al., 2011. Electrical capacitance tomography as a monitoring tool for high-shear mixing and granulation. Chem. Eng. Sci. 66 (18), 4090–4100.

Ritala, M., Holm, P., Schaefer, T., Kristensen, H.G., 1988. Influence of liquid bonding strength on power consumption during granulation in a high shear mixer. Drug Dev. Ind. Pharm. 14 (8), 1041–1060.

Rowe, R.C., Parker, M.D., 1994. Mixer torque rheometry: an update. Pharm. Technol. 18 (3), 74–74.

Saito, Y., Fan, X., Ingram, A., Seville, J.P.K., 2011. A new approach to high-shear mixer granulation using positron emission particle tracking. Chem. Eng. Sci. 66 (4), 563–569.

Sakr, W.F., Ibrahim, M.A., Alanazi, F.K., Sakr, A.A., 2012. Upgrading wet granulation monitoring from hand squeeze test to mixing torque rheometry. Saudi. Pharm. J. 20 (1), 9–19.

Sato, Y., Okamoto, T., Watano, S., 2005. Scale-up of high shear granulation based on agitation power. Chem. Pharm. Bull. (Tokyo) 53 (12), 1547–1550.

Schaeffer, T., Bak, H., Jaegerskou, A., 1986. Granulation in different types of high speed mixers. Part 1: effects of process variables and up-scaling. Pharm. Ind. 48, 1083–1089.

Schaeffer, T., Bak, H., Jaegerskou, A., Kristensen, H., 1987. Granulation in different types of high speed mixers. part 2: comparison between mixers. Pharm. Ind. 49, 297–305.

Shi, L., Feng, Y., Sun, C.C., 2011. Massing in high shear wet granulation can simultaneously improve powder flow and deteriorate powder compaction: a double-edged sword. Eur. J. Pharm. Sci. 43 (1–2), 50–56.

Sochon, R.P.J., Zomer, S., Cartwright, J.J., et al., 2010. The variability of pharmaceutical granulation. Chem. Eng. J. 164 (2–3), 285–291.

Stamm, A., Paris, L., 1985. Influence of technological factors on the optimal granulation liquid requirement measured by power consumption. Drug Dev. Ind. Pharm. 11 (2–3), 333–360.

Talu, I., Tardos, G.I., Ruud Van Ommen, J., et al., 2001. Use of stress fluctuations to monitor wet granulation of powders. Powder Technol. 117 (1–2), 149–162.

Terashita, K., Watano, S., Miyuanami, K., 1990. Determination of end-point by frequency analysis of power consumption in agitation granulation. Chem. Pharm. Bull. (Tokyo) 38 (11), 3120–3123.

Timko, R., Barrett, J., McHugh, P., et al., 1987. Use of a motor load analyzer to monitor the granulation process in a high intensity mixers. Drug Dev. Ind. Pharm. 13, 405–435.

Wade, J.B., Martin, G.P., Long, D.F., 2015. Controlling granule size through breakage in a novel reverse-phase wet granulation process; the effect of impeller speed and binder liquid viscosity. Int. J. Pharm. 478, 439–446.

Watano, S., 2001. Direct control of wet granulation processes by image processing system. Powder Technol. 117 (1–2), 163–172.

Watano, S., Tanaka, T., Miyanami, K., 1995. A method for process monitoring and determination of operational end-point by frequency analysis of power consumption in agitation granulation. Adv. Powder Technol. 6 (2), 91–102.

Watano, S., Sato, Y., Miyanami, K., 1997. Application of a neural network to granulation scale-up. Powder Technol. 90 (2), 153–159.

Watano, S., Numa, T., Miyanami, K., Osako, Y., 2001. A fuzzy control system of high shear granulation using image processing. Powder Technol. 115 (2), 124–130.

Werani, J., 1988. Production experience with end point control. Acta Pharm. Suec. 25, 247–266.

Whitaker, M., Baker, G.R., Westrup, J., et al., 2000. Application of acoustic emission to the monitoring and end point determination of a high shear granulation process. Int. J. Pharm. 205 (1–2), 79–92.

Wikström, H., Marsac, P.J., Taylor, L.S., 2005. In-line monitoring of hydrate formation during wet granulation using Raman spectroscopy. J. Pharm. Sci. 94 (1), 209–219.

Woyna-Orlewicz, K., Jachowicz, R., 2011. Analysis of wet granulation process with Plackett-Burman design—case study. Acta Pol. Pharm. Drug Res. 68 (5), 725–733.

Zega, J., Lee, D., Shiloach, A., et al., 1995. Scale-up of the wet granulation process for a dicalcium phosphate formulation using impeller power consumption. Poster, AAPS General Meeting.

Zlokarnik, M., 2006. Scale-up in Chemical Engineering, second ed Wiley-VCH Verlag GmbH & Co., Weinheim.

Zlokarnik, M., 1991. Dimensional Analysis and Scale-Up in Chemical Engineering. Springer Verlag, Berlin, Heidelberg.

INDEX

Note: Page numbers followed by "*f*" refer to figures.

Printed in the United States
By Bookmasters